Insect Species Conservation

Insects are the most diverse and abundant animals that share our world, and conservation initiatives are increasingly needed and being implemented globally, to safeguard the wealth of individual species. This book provides sufficient background information, illustrated by examples from many parts of the world, to enable more confident and efficient progress towards the conservation of these ecologically indispensable animals. Writing for graduate students, academic researchers and professionals, Tim New describes the major ingredients of insect species management and conservation, and how these may be integrated into effective practical management and recovery plans.

TIM NEW is Professor of Zoology at La Trobe University, Australia. He has broad interests in insect ecology, conservation and systematics, and has published extensively in these fields. He is recognised as one of the leading advocates for insect conservation. He is currently editor-in-chief of the *Journal of Insect Conservation*.

ECOLOGY, BIODIVERSITY AND CONSERVATION

The world's biological diversity faces unprecedented threats. The urgent challenge facing the concerned biologist is to understand ecological processes well enough to maintain their functioning in the face of the pressures resulting from human population growth. Those concerned with the conservation of biodiversity and with restoration also need to be acquainted with the political, social, historical, economic and legal frameworks within which ecological and conservation practice must be developed. The new Ecology, Biodiversity, and Conservation series will present balanced, comprehensive, up-to-date, and critical reviews of selected topics within the sciences of ecology and conservation biology, both botanical and zoological, and both 'pure' and 'applied'. It is aimed at advanced final-year undergraduates, graduate students, researchers, and university teachers, as well as ecologists and conservationists in industry, government and the voluntary sectors. The series encompasses a wide range of approaches and scales (spatial, temporal, and taxonomic), including quantitative, theoretical, population, community, ecosystem, landscape, historical, experimental, behavioural and evolutionary studies. The emphasis is on science related to the real world of plants and animals rather than on purely theoretical abstractions and mathematical models. Books in this series will, wherever possible, consider issues from a broad perspective. Some books will challenge existing paradigms and present new ecological concepts, empirical or theoretical models, and testable hypotheses. Other books will explore new approaches and present syntheses on topics of ecological importance.

The Ecology of Phytoplankton
C. S. Reynolds

Invertebrate Conservation and Agricultural Ecosystems
T. R. New

Risks and Decisions for Conservation and Environmental Management
Mark Burgman

Nonequilibrium Ecology
Klaus Rohde

Ecology of Populations
Esa Ranta, Veijo Kaitala and Per Lundberg

Ecology and Control of Introduced Plants
Judith H. Myers, Dawn Bazely

Systematic Conservation Planning
Chris Margules, Sahotra Sarkar

Assessing the Conservation Value of Fresh Waters
Phil Boon, Cathy Pringle

Bird Conservation and Agriculture
Jeremy D. Wilson, Andrew D. Evans, Philip V. Grice

Large Scale Landscape Experiments
David B. Lindenmayer

Insect Species Conservation
T. R. New

Insect Species Conservation

T. R. NEW

Department of Zoology, La Trobe University, Australia

CAMBRIDGE
UNIVERSITY PRESS

CAMBRIDGE UNIVERSITY PRESS
Cambridge, New York, Melbourne, Madrid, Cape Town, Singapore, São Paulo, Delhi

Cambridge University Press
The Edinburgh Building, Cambridge CB2 8RU, UK

Published in the United States of America by Cambridge University Press, New York

www.cambridge.org
Information on this title: www.cambridge.org/9780521510776

First published 2009

Printed in the United Kingdom at the University Press, Cambridge

A catalogue record for this publication is available from the British Library

Library of Congress Cataloging in Publication data
New, T. R.
Insect species conservation / T. R. New.
    p.   cm.
Includes bibliographical references and index.
ISBN 978-0-521-51077-6 (hardback)
1. Rare insects.   2. Insects – Conservation.   I. Title.
QL467.8.N49   2009
639.9′757 – dc22   2009007353

ISBN 978-0-521-51077-6 hardback
ISBN 978-0-521-73276-5 paperback

Contents

Preface

This short book is about conserving insects, the most diverse and abundant animals that share our world. In particular, it is about the common focus of conserving individual species of insects. This so-called 'fine filter' (or 'fine grain') level of conservation parallels much conservation effort for better-understood groups of animals such as mammals and birds, for which species-focused conservation exercises are commonplace. The need for insect conservation can appear puzzling, and how to undertake it can seem daunting to the many conservation practitioners unfamiliar with insects but to whom vertebrates or vascular plants are familiar – and, thus, that they can treat with greater confidence because of being within their range of practical expertise. We are thus dealing with insects as specific targets or individual foci for conservation. My main aim is to provide sufficient background information, illustrated by examples of insect species needs and conservation programmes from many parts of the world, to enable more confident and efficient progress for conservation of these ecologically indispensable animals. I hope to demonstrate and clarify to potential managers what the major ingredients of insect species management for conservation may be, and how those needs and ingredients may be integrated into effective and practical management or recovery plans.

The examples demonstrate the great variety of needs of ecologically specialised insects, the small scales over which they may operate, and how both assessment of conservation status and design of species conservation necessarily differs from that for many of the more popular and more widely understood organisms.

The need for such an appraisal has been stimulated largely by my experiences in Australia, where most people involved 'officially' in managing insects for conservation, such as by belonging to State or Territory conservation or related agencies, are (in common with many people in similar positions elsewhere in the world) not primarily entomologists, but versed in the management or ecology of vertebrates or other organisms.

They commonly fail to appreciate the idiosyncrasies and importance of the threatened insect species with which they are obliged to deal. Similar perspectives are also common elsewhere, but this book is also an opportunity to present some Australian cases to a wider readership and to integrate them with better-known examples from elsewhere to provide a wide geographical picture of progress in insect species conservation. Much of the relevance of Australian cases in this perspective reflects the relatively recent rise of insect conservation interest in the country, in contrast to its much longer recognition in much of the northern hemisphere, and that it has thus been able to draw on the much more substantial framework of insect conservation practice established elsewhere. I emphasise that these cases are not presented as examples of 'best practice', but simply as ones with which I am most familiar, and that are sufficiently varied to demonstrate successes and failures of various components of insect species management.

The book deals primarily with insect ecology and its central role in understanding and formulating practical conservation measures, and also with the legislative and regulatory environment relevant to insect conservation at this level. It is not a compendium of sampling theory and methods. Those are available elsewhere (see, for example, the books by Southwood & Henderson 2000; New 1998; Samways *et al.* 2009), but references to various methods used for sampling and monitoring are inevitable and the above texts may be consulted for further details of these. Much of the best insect conservation practice hangs on the approaches and field methods employed. Many individual species studies contain original, often innovative, modifications of standard methods tailored to the biology of the focal species, and the 'methods' section of published papers and reports usually bears close scrutiny. Likewise, many of the broader aspects of insect conservation biology are included in the volume arising from a recent Royal Entomological Society symposium on this topic (Stewart *et al.* 2007). Rather than revisit all those themes, I discuss insect biology as the scientific background to insect species conservation, the scope and extent of species conservation, and how the requisite management may be undertaken effectively through realistic planning and regulation justified by biological understanding. My main emphasis is on the design and implementation of effective insect species management plans.

'Species level conservation' is the means through which many people have been introduced to insect conservation and to the often intricate conservation needs of specialised insect species, with the important lesson that every insect species differs in subtle ways from every other, and that it is often unwise to extrapolate uncritically ecological details from one species even to its closest relatives. Nevertheless, each of the many individual species management plans which have beeen published demonstrates principles, ideas and − sometimes − detail that can help refine plans for other species.

I do not deal in this book with the 'coarse filter' levels of insect conservation, namely insect assemblages and communities, despite the increasing needs for these, and the accelerating realisation that they may be the only practical way for insect conservation to proceed effectively in many parts of the world. This wider need occurs simply because the vast number of individual needy species is overwhelming. They cannot all be given individual attention, and some form of allocating priority or triage between deserving species is inevitable, with the consequence that many needy species will be neglected. Those wider levels of focus, emphasising the conservation of insect diversity, are summarised admirably by Samways (2005). Nevertheless, understanding the ecological peculiarities and details of individual insect species' conservation needs will continue to emphasise their importance as flagships for the less-heralded components of the world's biodiversity, and to enhance understanding of the natural world. The lessons learned from insect species conservation programmes over the past half century, in particular, provide important leads toward promoting more efficient and more effective programmes for the future. Accelerating that aim is a main driver of this book.

In many parts of the world, resources available for insect species conservation are in very short supply, and their allocation for best effect difficult to arrange or, even, to suggest. Resident concerned entomologists or conservation biologists are few over much of the tropics, for example. The wellbeing of individual butterflies, dragonflies or beetles (or, even less so, of barklice or flies) is understandably accorded very low priority in relation to pressing requirements of human welfare and in places where land use for food production for people is a primary need. Much of this book is based on examples from countries where this is not the case, and where such aspects of conservation (some of them based on many decades of experience and very detailed planning, and well-resourced interest and management) are accepted easily as part of a 'national psyche'. In particular, I draw on selected examples from Europe (in particular the United Kingdom), North America, Australia, New Zealand, Japan and South Africa to discuss the development of insect species conservation practice and theory. Essentially, these are predominantly from the temperate regions of the world, and equivalent species conservation programmes in much of the tropics simply do not exist, other than by rare chance. In all these named regions, individual species cases have been central to development and promotion of insect conservation interests. Many of them are based on 'charismatic' insects, particularly butterflies, dragonflies and some larger beetles, that have captured public interest in various ways, and some of which have become significant local flagships for wider conservation efforts. A broad spectrum of priorities and tactics for conservation collectively contribute to a synthesis, which may lead toward more effective protocols for wider adoption. At the least, wider awareness of the varied approaches, activities and possibilities, many of them intermeshing excellent science with protective regulation or legislation, should enable managers to aid the future of many insect species through improving practical conservation, and also to assess how insect species conservation programmes may participate in assuring wider benefits and be pursued with greater confidence.

Some cases are discussed in greater detail, and a selection are presented in Boxes in the text, to illustrate particular management points or approaches to study or assessment. Collectively these provide examples of recovery measures that have worked, or have been unsuccessful, and indicate the kinds of information and practice that may contribute to the eventual outcome. Some will be well known to entomologists as 'classics' of insect conservation but, equally, they will commonly be less familiar to other people – except, perhaps, through casual acquaintance. They provide the foundation both for wider understanding and the lessons learned so far in a rapidly evolving science, and also for energetic debate about optimal ways to proceed and develop what we understand at present to ensure a more secure future for insects in the increasingly unnatural world.

Acknowledgments

The following agencies and publishers are thanked for permission to use or adapt previously published material in this book: AULA-Verlag GmbH, Wiebelsheim; Australian Government Department of the Environment, Water, Heritage and the Arts, Canberra; Blackwell Publishing, Oxford; Butterfly Conservation, Wareham; Czech Academy of Sciences, Ceske Budejovice (European Journal of Entomology); Department of Conservation, Wellington, New Zealand; Elsevier Science; Island Press, Washington, DC; Minnesota Agricultural Experiment Station, University of Minnesota; Oxford University Press, Oxford; Springer Science and Businesss Media, Dordrecht; Surrey Beatty & Sons Pty Ltd, Chipping Norton, New South Wales. Individual references to all sources are given in legends to figures and tables. Every effort has been made to obtain permissions for such use, and the publishers would welcome news of any inadvertent oversights.

I greatly appreciate the continuing interest and support of Dominic Lewis at Cambridge University Press during the gestation of this book. Chris Miller guided it through production and I also thank Lynn Davy for careful copy-editing.

1 · *Needs and priorities for insect species conservation*

Introduction: extinctions and conservation need

Vast numbers of insect species exist on Earth. They are the predominant components of animal species richness in most terrestrial and fresh-water environments. and by far outnumber many more familiar or popular animal groups, such as vertebrates. Estimates of the numbers of living insect species range up to several tens of millions; no one knows how many, but biologists accept easily estimates within the range of 5–10 million species as realistic. However, only about a million insect species have been formally described and named. The very levels of uncertainty over numbers of existing insect species are sobering reminders of what we do not know of our natural world. They help to emphasise our general ignorance over the diversity and ecological roles of many of the organisms that drive and maintain the ecological processes that sustain natural communities.

There is little doubt that very many insects have declined over the past century or so in response to human activities in many parts of the world. Such losses, reflecting changes we have made to their habitats and the resources on which they depend, have been documented most fully (but still with many substantial gaps in knowledge) in some temperate regions of the world (Stewart & New 2007). Insect extinctions and declines may be considerably greater in much of the tropics (Lewis & Basset 2007), where they are less heralded, but where numbers of insect species appear to be vastly higher than in many temperate regions. Unlike most groups of vertebrates, for which extinctions have sometimes been documented in considerable detail, extinctions of most insects have not been described – other than, predominantly, for a few Lepidoptera in northern temperate regions. Indeed, more than half of the recently documented insect extinctions are Lepidoptera. Many of the problems of determining the fact and likelihood of recent insect extinctions were explored by Dunn (2005), who suggested that insects might be especially prone to two forms of extinction that are rare in other taxa: extinctions of narrow habitat specialists, and coextinctions of species with their hosts, be these animals or plants. The two are to some extent parallel, and Dunn's conclusion that both these categories tend to be ignored by conservation programmes that focus on vertebrates or plants is relevant here. The second of his categories applies, for example, to insect parasitoids (p. 45) and ectoparasites, as well as to monophagous herbivores, many of which are among the cases noted in this book as causing major conservation concerns. Many intricate and obligate relationships are involved: thus, for the Singapore butterflies, Koh *et al.* (2004) suggested that many more butterflies are likely to become extinct along with their host plants, as they depend entirely on those particular host plants. Recent declines of pollinating insects in many parts of the world have caused concerns for the plants that depend on these. Again, such relationships may be very specific, and emphasise the intricacies of many of the ecological interactions in which insects participate. The need to hand-pollinate rare endemic plants in Hawaii and elsewhere demonstrate eloquently one category of the cascade effects that may flow from losses of ecologically specialised insects.

This lack of detailed knowledge of the extent of extinctions, however, cannot be allowed to lull us into false confidence that insects do not need attention to sustain them. In short, even though rather few global extinctions among recent insects have been confirmed (Mawdsley & Stork 1995), many insects are inferred strongly to have declined and are in need of conservation measures if they are to survive. Local extinctions of insects are frequent, and are the primary focus of much conservation. There are clear logistic limitations to the extent to which those deserving species can be treated individually, but attempts to do so have fundamentally increased our appreciation of insect biology and led to greatly improved conservation focus for species level conservation in many different contexts. These contexts range from the initial selection of candidates for consideration (and the criteria by which the 'most deserving species' may be given priority) to the effective design and implementation of management. In most parts of the world, even most of the insect species designated formally as 'threatened' have not become the subjects of focused species management plans. To some extent, this simply reflects the tyranny of large numbers of candidates and consequent impracticability of dealing with them, but this is often compounded by uncertainties over how to assess those priorities rationally and convincingly and, often, necessarily from an inadequate framework of biological knowledge and understanding.

At the outset, the conservation of insects (and of other invertebrates) reflects a number of features of scale that render them rather different from the vertebrates and higher plants generally more familiar to conservation managers, in addition to their taxonomic complexity, noted below. Some of these are noted here to aid perspective in conservation planning.

1. Most insects are small, and the normal population dynamics of many species is characterised by substantial intergenerational changes in numbers, so that detecting real trends in decline may necessitate observations over many generations.
2. Many conservation needs for insects arise from the focal species being extreme habitat specialists with very intricate resource requirements.
3. Many species have very narrow distributions in relation to those resources, with 'narrow range endemism' apparently a very common pattern. In most cases we do not know if narrow distributions are wholly natural, or represent declines from formerly broader ranges, for example as a result of habitat fragmentation.
4. Most insects have short generation times, with one–few generations a year being the most frequent patterns of development, but each insect species may have a largely predictable phenology within a given area. A univoltine (annual) life cycle implies a strongly seasonal pattern of development, so that differing resources for adult and immature stages must be available at particular times each year.
5. A corollary to this is that each life history stage may be available for inspection or monitoring only for a short period each year or each generation, with activity (essentially, opportunity for inspection) governed strongly by weather factors.
6. The suitability of a site for an insect depends not only on consumable resources but also on microclimate. Temperature is an important determinant of site suitability, so that attributes such as bare ground, vegetation cover and density, site aspect and slope may influence an insect's incidence and abundance in unexpectedly subtle ways.
7. Many insects are relatively immobile, so that they are predominantly restricted to particular sites or microhabitats. The factors that determine suitability of a microhabitat to an insect are commonly of little or no concern for other organisms. Likewise, very small sites (such as tiny patches of roadside vegetation) may be critically important for

particular insects but dismissed as too trivial to consider for vertebrates. Some insect species are known only from such minute areas,
of a hectare or less.

8. Many insects manifest a metapopulation structure (p. 91), itself not
always easy to define for species found in low numbers and widely dispersed, but of fundamental importance in estimating risk of extinction
and, in conjunction with 7 (above), the accessibility of microhabitats
in the wider landscape.

Planning priorities among species

Setting priorities for insect species conservation among an array of
acknowledgedly worthy candidates is indeed difficult. This is despite
species being the most tangible focus of practical conservation to many
people, as entities to which we can relate, in contrast to more nebulous
and complex entities such as communities and ecosystems. Formal listing of insects on a schedule of 'protected species' or 'threatened species'
commonly obliges some form of further investigation or action. Yet the
species which come to our notice as needing conservation attention, particularly when these are insects or other poorly documented organisms
without a strong body of public support, are simply the small tip of a very
large species iceberg. They are commonly simply those taxa over which
someone, somewhere, has concerns, and are not fully representative of
the greater needs of that group of organisms. Under Australia's federal
Environment Protection and Biodiversity Conservation Act (EPBC), for
example, and mirrored in all eight of the country's State or Territory
legislations, together with virtually all similar or parallel legislations elsewhere in the world, only a handful or so of the possible tens of thousands
of worthy candidate insects are scheduled at present, and numerous highly
diverse groups of invertebrates are entirely missing. There is considerable
bias in what invertebrates are listed, or can be included in such lists –
or, perhaps, even that should be listed in this way, for two main reasons:
(1) our lack of capability and resources to deal practically with large
numbers of species to which we become committed, and (2) lack of
rational bases based on sound information to designate the most 'deserving' species for our limited attention. It is no accident that a high proportion of insects listed in Australia, and elsewhere, are butterflies, the
most popular group of insects and ones that can be promoted effectively
as 'flagships' from a climate of sympathy for their wellbeing combined
with reasonably sound evidence, arising largely from the concerns of

hobbyists, of decline and conservation need. Thus the Bathurst copper butterfly (*Paralucia spinifera*) was the first invertebrate to be classified as 'endangered' in New South Wales, and its close relative the Eltham copper (*P. pyrodiscus lucida*) was amongst the first invertebrates listed formally in Victoria. Consequent concern for both these taxa has done much to promote wider interest in insect conservation in both states, with *P. spinifera* becoming the first butterfly to be listed federally in Australia, as 'vulnerable'. But this bias towards insects for which public sympathy is evident does not mean that others are of no concern; simply that listing many a psocid or small fly (in any part of the world) would serve little practical purpose other than conveying some slight message of political need – but, broadly, also likely to evoke a certain amount of ridicule. However, one outcome is clearly a strong bias toward favoured groups, although these groups need not necessarily be those that are well known. In an early survey of which insects in Australia might be threatened, Hill and Michaelis (1988) sought feedback from a wide circle of informed correspondents. One outcome of the exercise was that 56 of the 62 Diptera nominated were from the family Drosophilidae, representing the zeal of a single specialist. From that list, it could be inferred that no others of the hundred or so Australian families of flies are of concern, an inference that may be very misleading and which might not be apparent to politicians and managers relying on such lists for setting their priorities.

For many groups of insects, even in the best-documented regional faunas, taxonomic knowledge is still very incomplete, and our ability even to recognise species consistently is very limited. Within groups such as parasitoid Hymenoptera, for example, numerous complex suites of 'biological species' occur, differing in fundamental biological characteristics such as host specificity but indistinguishable on morphological features (Shaw & Hochberg 2001). As in numerous other insect groups, the proportion of species of these generally tiny wasps yet recognised may be of the order of only 10%–20% globally. Even fewer of these have been named scientifically. Up-to-date handbooks or other identification guides for many insect groups simply do not exist and, without ready access to a large and well-curated reference collection, a non-specialist has little chance of identifiying most of the species encountered. Even then, or with the best possible advice from acknowledged specialists, many problems of recognition and identification will persist.

Thus, discovering new, undescribed or undiagnosable insect species is a routine (and, commonly, unintentional) activity for entomologists taking up the study of almost any insect order or other group, particularly in the tropics or southern temperate regions of the world. In contrast, discovery of a new large vertebrate, particularly a mammal or bird, is a much rarer and more newsworthy event, as a consequence of much fuller early documentation of these animals, their wide interest and popular appeal, and their comparatively low diversity. Whereas virtually all species of animals such as mammals and birds have been recognised and assessed reliably for conservation need and allocation of conservation status, equivalent capability and coverage has not been achieved for insects, and such comprehensive assessment is utopian. Even for the best-known insect groups, many gaps remain. In addition, the real conservation status of many of the species that *are* included on protected species lists and the similar schedules that may accord them conservation priority remains controversial, with their practical conservation needs inferred rather than scientifically unambiguous. Many species listings have not been reviewed critically, and the species' status and needs may have changed considerably since it was originally signalled for conservation need.

In short, most lists of 'threatened' insects and other invertebrates tend to be either too short to be fully representative or too long for us to be able to cope with responsibly or comprehensively. Some level of selection or triage is almost inevitable in developing preliminary or idealistic 'lists' to 'practical conservation', with numerous species admitted as deserving and needing conservation neglected simply because our resources cannot cope with all of them. This problem is not peculiar to insects, of course, but the sheer numbers involved make the problem massive and very obvious. It follows that grounds for placing any insect species on such a list, and for later selecting it further for attention, must be sound, clearly understood, and responsible, as a foundation for committing effort and resources to its conservation based on credibility. It is also a step by which other species are likely to be deprived of any equivalent management. Many insects on such schedules are not necessarily threatened; and many threatened species are not necessarily listed. Referring to parasitoid Hymenoptera, Hochberg (2000) summarised the impracticability of species level conservation efforts by writing 'Their staggering diversity simply means that the focused conservation of, say, 1000 individual species over the next century may be numerically infinitesimal compared to the actual number of endangered species'.

This reality can appear overwhelming. Nevertheless, there is clear need to promote and undertake insect conservation on several levels, although many practitioners have emphasised the inadequacies of species level conservation as a mainstay procedure. However, and as noted earlier, 'people relate to species', and insect species are undoubtedly important both in their own right and as vehicles through which the massive variety of insect ecology and its relationships to wider conservation of natural processes and ecosystems can be advertised and displayed.

It is thus important that conservation cases for the insect species we *do* select for individual conservation attention are presented on grounds that are convincing and objective, preferably measured against criteria that are accepted readily as suitable. Two main contexts of formally assessing conservation need may sometimes become confused: first, to provide an absolute statement of conservation status of the species, and, second, to rank that species for relative priority within the context of a local fauna or taxonomic group. In short, we need to determine (1) whether the insect species is threatened with extinction and, if so, how, and (2) the grounds for giving it priority for attention over other deserving threatened species. We are faced with both an absolute decision (is the insect threatened or not threatened?) and a relative decision for ranking (is it more or less 'deserving' than others found in the same higher taxon, biotope or area?). These determinations give our cases credibility. With that assured, each insect species selected for individual conservation attention can contribute to wider advocacy and understanding for the importance of invertebrate conservation, at any scale.

Criteria for assessing priority

Each of these contexts requires objective appraisal against some defined criterion/criteria, with the usual outcome being some form of advisory or legislative categorisation in the form of a list of threatened or protected species: broadly 'listing', ideally accompanied by assessment of priority or urgency of conservation need. In some cases, the criteria for categorising and listing insects are determined formally; in others they are a working guide. Table 1.1 exemplifies the variety of features that may be considered, in that case for European dragonflies (van Tol & Verdonk 1988). Formal criteria for listing insects differ substantially under various legislations, so that comparison of lists from different agencies and places and made at different times is difficult. Many of the assessment criteria are based on the IUCN categories of threat of extinction, historically mainly those promoted in 1994 but more recently revised (IUCN 2001), and such appraisals have become major drivers of assessing conservation status. However, many entomologists have found inadequacies and frustrations in trying to apply them to insects with confidence,

Table 1.1 *An example of a set of criteria used for ranking insect species in lists of threatened taxa: the parameters used for Odonata in Europe by van Tol and Verdonk (1988)*

Parameter	Likelihood of inclusion	
	Lower	Higher
Intraspecific variation	small	large
Species range	large	small
Position of Europe in species range	edge	centre
Species endemic to Europe	no	yes
Population density	high	low
Population trend: twentieth century	increase	decline
Trophic level of biotopes frequented	eutrophic	oligotrophic
Habitat range	eurytopic	stenotopic
Resilience to environmental changes	high	low
Dispersal power	high	low
Potential population growth	high	low
Ecological strategy	*r*-strategist	*K*-strategist
Conspicuousness	small	large
Effect of construction of artificial biotopes	high	small

and numerous modifications have been made in attempts to reduce the reliance on quantitative thresholds of extinction risk that are rarely, if ever, available for insects, and are commonly compounded by lack of knowledge of population structure. Clarke and Spier (2003) applied a form of analytical software (RAMAS Red ®: Akçakaya & Ferson 1999) to available population data for a variety of Australian invertebrates, but the considerable uncertainty involved led to polarisation between the categories of 'Data Deficient' and 'Critically Endangered'.

The IUCN criteria, displayed in Fig. 1.1, have central importance as an avenue to assessing conservation status, but include criteria that are difficult or impossible to apply to most insects. In particular, any data available for determining quantitative thresholds of population decline and likelihood of extinction involve considerable speculation, and usually cannot be employed with confidence for such poorly documented taxa. Nevertheless, some such categorisation reflecting urgency of need (so that, in the IUCN categories, 'Critically Endangered' ranks above 'Endangered' and this in turn ranks above 'Vulnerable' in a hierarchy of threat categories) is important in allocating priority on grounds of

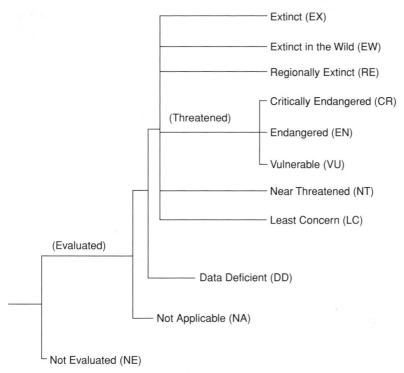

Fig. 1.1. Schematic representation of the IUCN Red List Categories (IUCN 2001). 'Threatened' includes the categories of 'Critically Endangered', 'Endangered' and 'Vulnerable'.

threat intensity. Guidelines to using the recent IUCN categories (IUCN 2003) note, as is commonly not acknowledged or appreciated elsewhere, that 'assessment of extinction risk' and 'setting conservation priority' are related but different processes. The former usually precedes the latter, which can also incorporate a variety of other considerations. Whatever the criteria used, the placing of an insect on any advisory or regulatory list of threatened or protected species must be a responsible action, with the grounds for doing so transparent and justifiable. Subsequent ranking for conservation attention is likely to involve a further round of triage, and neglect of the 'less worthy' species simply because they are ranked lower in a climate of limited support for action. At the least, including an insect species on any such list is likely to promote it for conservation attention over non-listed species, and may be a politically expedient action in indicating need for support.

Box 1.1 · *The IUCN categories of threat to species and the criteria used to designate these formally (after IUCN 2001)*

The following summary table emphasises the quantitative differences between the three categories of 'critically endangered' (CR), 'endangered' (E) and 'vulnerable' (Vu). Note that these data are rarely, if ever, available for insects, but the criteria serve to reflect the nature of differences between these designations of risk of extinction, which form the foundation of much categorisation of conservation status of species in legislation.

	Critically endangered	Endangered	Vulnerable
A *Declining population*			
Population declining at a rate of ... using either	>80% in 10 years or 3 generations	>50% in 10 years or 3 generations	>20% in 10 years or 3 generations
1. Population reduction estimated, inferred or suspected in the past OR			
2. Population decline suspected or projected in the future, based on direct observation, an abundance index, decline of habitat, changes in exploitation, competitors, pathogens, etc.			
B *Small distribution*			
Either extent of occurrence ...	$<100 \, \text{km}^2$	$<5000 \, \text{km}^2$	$<20\,000 \, \text{km}^2$
OR			
Area of occupancy ... and 2 of the following 3:	$<10 \, \text{km}^2$	$<500 \, \text{km}^2$	$<2000 \, \text{km}^2$
1. Either severely fragmented or known to exist at a number of locations			

	Critically endangered	Endangered	Vulnerable
2. Continuing decline in habitat, locations, subpopulations or mature individuals			
3. Fluctuations of more than one order of magnitude in extent, area, locations or mature individuals.			
C *Small population size and decline*			
Number of mature individuals . . . and 1 of the following 2:	<250	<2500	<10 000
1. Rapid decline of . . .	>25% in 3 years or 1 generation	> 20% in 5 years or 2 generations	> 10% in 10 years or 3 generations
2. Continuing decline of any rate and Either:			
Populations fragmented with . . .	All subpopulations <50	All subpopulations <250	All subpopulations <1000
or All individuals in a single population			
D *Very small or restricted population*			
Number of mature individuals . . .	<50	<250	<1000 Or area of occupancy <1000 km^2, or locations <5
E *Quantitative analysis*			
Risk of extinction in the wild	>50% in 10 years or 3 generations	>20% in 20 years or 5 generations	>10% in 100 years

As suggested above, 'extinction risk' is only one of the criteria, albeit an important one emphasised by many workers, that may accord a species priority for conservation attention. As Miller *et al.* (2007) emphasised, the intent of IUCN categorisation is indeed to evaluate extinction risk, and not to prioritise species for conservation. The latter process must incorporate a much wider array of factors, and allow for weighting of these in any individual context. One example is noted in Box 1.2, but the major categories of factors for conservation priority listed by Miller *et al.* are extinction risk, distributional factors, biological factors, societal values, logistical factors, economic factors, and 'other factors'.

Box 1.2 · *Setting priorities for conserving insect species within a group: a model for Irish bees*

Approaches to setting priorities for conservation within a group of threatened species continue to develop, with varying degrees of prag- matism and reality. Within a region, such priorities tend to reflect (1) national status, as revealed by a Red Data listing or similar appraisal, (2) international status and importance, and (3) global and interna- tional considerations. Each of these can be ranked for priority broadly as 'low', 'medium' or 'high'. Using the Irish bees (of which 30 of 100 species were categorised as threatened at the commencement of the survey), Fitzpatrick *et al.* (2007) sought to produce a national list of species of conservation priority. They employed eight criteria, draw- ing on existing documentation, and several assumptions underpinned the survey, so that the national list of priority species should:

1. Be based on the national red list.
2. Take into account the significance of the national populations at the global or continental level.
3. Take into account the conservation status at the global, continental and regional level.
4. Take into account key biological, economic and societal factors.
5. Be compatible with existing conservation agreements and legisla- tion.
6. Be simple so that it can be easily understood and updated.

The eight criteria also depended on some prior knowledge of status from a national red data list and categorisation under the IUCN criteria of threat, and form an operational sequence, as follows:

1. Global or European conservation status. Place internationally threatened species (those on global or European red lists) directly on the list of national priority species, regardless of Irish red list status.
2. Species protected by existing national agreement or legislation. These are also placed directly on the list, as above.
3. International importance. Place internationally important species directly on the list, as above. Defined in this individual context as species with >20% of global or continental populations within Ireland, recognising that this percentage may need adjustment elsewhere (for example, in relation to areas involved).
4. European conservation status, for taxa for which no published summary of status is available. Estimates of threat were made by considering their status in each of five divisions of Europe, and including regionally extinct taxa. Species included are those threatened in Ireland and regionally extinct or threatened in at least 50% of regions across their range.
5. Regional conservation status, involving consideration of the species' conservation status within its own region. Included species are those threatened in Ireland and in at least 50% of countries within the region.
6. Keystone species. Emphasises ecological importance, as species whose loss may lead to secondary extinctions (such as by being the sole pollinators of particular plants).
7. Species of recognised economic value. All threatened species on national red lists that have recognised economic value were transferred to the priority list, as above.
8. Species of cultural importance. All threatened species on the national red list that are of cultural importance (for example, as a national emblem) at the national level were transferred to the priority list, as above.

This approach acknowledges both biological and non-biological factors in according an insect priority. Species on the final list may then be treated individually with priority among them reflecting a variety of factors. Fitzpatrick *et al.* (2007) listed the following influences relevant at this stage: public appeal of the species, educational values, flagship species status, type of action needed, feasibility, urgency (driven by risk of extinction), conflicting issues, cost of action, economic loss if protected, involvement of non-government organisations.

Ranking within a local fauna, rather than absolute categorisation of status, may demand a somewhat broader approach for satisfactory comparison amongst the species involved. Additional ranking criteria to those based on threat can include the extent to which the species represents an isolated taxonomic lineage. Thus the tiny, locally endemic Australian damselfly *Hemiphlebia mirabilis* has long been considered to be a basal taxon within the order Odonata, and is currently the only extant member of a superfamily (Hemiphlebioidea). It thus has no close relatives, and its loss would represent loss of an entire ancient lineage, of considerable evolutionary significance. On such grounds, for which there are numerous parallels within the insects, it could be deemed more 'valuable' than an equally threatened species with many close relatives – for example, one that represents a genus containing many closely related species. This aspect of priority among taxa was discussed by Vane-Wright *et al.* (1991).

For butterflies of Europe (van Swaay & Warren 1999), emulated later for Australia (Sands & New 2002; see p. 29), initial decisions involved determining the level of 'global importance' of the species, initially by determining whether each species was restricted to, or even resident in, the area under appraisal. The latter context arose, for example, in assessing if some of the butterflies recorded as adults from islands in the Torres Strait (separating Australia from New Guinea) were simply vagrant from New Guinea, or whether resident populations occurred there. Van Swaay and Warren (1999) used a ranking based on 'SPEC' (SPecies of European Concern), with four main categories as follows.

SPEC 1. Species of global conservation concern, restricted to Europe and considered globally threatened. These species are of the highest conservation concern, and require conservation wherever they occur.
SPEC 2. Species concentrated in Europe and threatened in Europe.
SPEC 3. Species threatened in Europe, but with headquarters both within and outside Europe.
SPEC 4a. Species that are European endemics but not considered threatened at present.
SPEC 4b. Species with global distribution concentrated in Europe but not considered threatened at present.

Red Data Lists and Red Data Books are advisory documents, highlighting the plight of organisms based on application of the IUCN categories of threat or a reasoned alternative, and summarising conservation needs to varying extents. Two global compendia, both now somewhat outdated but remaining fundamental reading for insect conservationists throughout the world, are for invertebrates (Wells *et al.* 1983) and swallowtail butterflies (Collins & Morris 1985), the latter highlighting one of the most charismatic and popular groups of insects. The first was of massive importance in, for the first time, displaying the variety of invertebrate life and its conservation needs to the world's conservationist fraternity. Wells *et al.* included accounts of an array of individual species, from many parts of the world and many taxonomic groups, to exemplify the wide range of invertebrate conservation needs. Collins and Morris (1985), in contrast, surveyed all members of one butterfly family, the 573 species of Papilionidae. Of these, 78 species were considered to be threatened in some way, or simply 'rare'. In this particular context, 'rare' refers to 'taxa with small world populations that are not at present Endangered or Vulnerable, but are at risk' (Collins & Morris 1985, p. 3). A variety of national Red Data Books or other summaries of threatened insect species needs have been published since then, of varying completeness and complexity, and with some variations from the IUCN assessment categories to accommodate more local needs. However, for many of the insect species listed by Wells *et al.*, little has occurred since then to focus their needs more completely. Thus, Wells *et al.* (1983) noted two species of Australian torrent midge (Diptera: Blepharoceridae) as endangered because of their vulnerability to particular developments involving river/stream regulation and change, but these species have not been examined critically by additional extensive fieldwork since then. The giant torrent midge (*Edwardsina gigantea*) was subsequently listed as 'endangered' by IUCN; Clarke and Spier (2003) suggested that it might be 'critically endangered'. This evaluation resulted from its very specialised environmental needs (restricted to clear torrential streams in cool mountain areas over part of the Great Dividing Range) rendering it vulnerable to any changes from pollution or water regulation, coupled with very weak flight capability. Likewise, many of the species of Papilionidae noted as threatened by Collins and Morris (1985), some of them discussed further by New and Collins (1991), have not yet become the focus of individual recovery efforts. This reflects that many of these spectacular butterflies occur in remote areas of the tropics where other issues must prevail, but also that no new information has been accumulated to clarify their needs and status further.

The British 'Insect Red Data Book' (Shirt 1987) includes summaries of eight orders, some of them treated very incompletely because only certain subsections of most large orders lend themselves to such assessment. Even for Lepidoptera, treatment is necessarily very uneven, with the butterflies documented more fully than the larger moths, and many of the smaller moths (the 'microlepidoptera', by far the most diverse group) very poorly known. Specific conservation recommendations are included for many of the species listed; although not comprehensive, these are invaluable pointers to key aspects of need. As examples, for the moth *Phyllocnistis xenia* the text reads 'The grey poplars on which the larvae feed require protection'; and for the beetle *Crytophagus falcozi* 'The preservation of ancient beech and oak in Windsor Forest'. Comments for some other included species are more general, such as on site status: for the stratiomyid fly *Oxycera dives*, Shirt (1987, p. 305) records 'Whilst the three most recent sites are all within SSSIs [=Sites of Special Scientific Interest], the habitat is so small and fragile that Vulnerable status is justified'. Simple statements on status and threat accompany other species in this account.

The insects (and other species) noted in advisory compilations for a group or region are commonly those first selected for conservation priority treatment within that region, as a consequence of (1) this prior signal of threat and conservation significance, and (2) that the same people are involved in both initial appraisal and wishing to pursue recovery measures. In effect, Red Lists and the like commonly constitute *de facto* conservation priority lists, even though they are mainly compiled with the very different purpose of categorising risk of extinction. Initially, Red Lists were designed to assess this risk over the entire species over the global range. More local lists essentially represent a 'regional subpopulation' (Gärdenfors 1996) with possibly inflated external risks from those accorded the species when appraised across its entire range. This problem has been addressed by preparation of regional guidelines for using the IUCN categories (IUCN 2003) and, although with little specific mention of invertebrates, the application of this system has been reviewed by Miller *et al.* (2007).

The absolute size of any insect populations needed to satisfy fully the IUCN criteria of threat will almost invariably be unknown. Despite this, it may be possible to infer or confirm a number of practical components of value, without seeking precise quantification: (1) that the population present is indeed resident and breeding; (2) whether it provides individuals to other populations in the vicinity, or receives immigrants from them: thus, is the population open, closed, or a metapopulation (p. 91); and (3) whether there is any trend evidence of decline in numbers or distribution. In some contexts, these or similar ideas have been incorporated into criteria for listing species. For example, under Australia's Australian Capital Territory (ACT) Nature Conservation Act, the criteria fulfilled for listing the golden sun–moth (p. 30) were given as:

1. The species is observed, estimated, inferred or suspected to be at risk of premature extinction in the ACT region in the near future, as demonstrated by:
    Current severe decline in population or distribution from evidence based on:
    Direct observation, including comparison of historical and current records: and
    Severe decline in quality or quantity of habitat.
1.2.5 Continuing decline or severe fragmentation in population, for species with a small current population. (ACT Government 1998)

As another example, moths were selected for inclusion on the UK Biodiversity Action Plan list of priority species on the following criteria:

1. Threatened endemic and other globally threatened species.
2. Species where the UK has more than 25% of the world or appropriate biogeographical population.
3. Species where number or range have declined by more than 25% in the past 25 years.
4. In some instances, where the species is found in fewer than 15 of the 10 km squares used as recording units in the UK.
5. Species listed in the European Union Birds or Habitat Directives, on the Bern, Bonn or CITES Conventions, or the Wildlife and Countryside Act 1981 or the Nature Conservation and Amenity Lands (Northern Ireland) Order 1985 (Parsons 2004).

Such trends can sometimes be inferred reliably without need for precise numerical data.

Some sensible compromises between 'strictly quantitative' and 'relatively quantitative' criteria are useful (for any organisms, not insects alone). The criteria used under Australia's federal act, for example (Table 1.2), include terms such as 'extremely low', 'very low' and 'low' for numbers of mature adults, with these estimates based on <50, <250 and <1000, respectively. However, the Advisory Committee does not apply these and other threshold criteria strictly but 'has regard to them when making judgments about species in terms of the biological contexts, and on a case-by-case basis'. The guidelines, criteria and threshold values are distributed with the nomination form for listing species.

For those insects that are confined to single small sites and disperse little, it may indeed be feasible to estimate population size by mark–release–recapture methods or direct counts. However, interpreting those

Table 1.2 *Criteria for listing species under Australia's Environment Protection and Biodiversity Conservation Act 1999*

| | Category | | |
| | Critically Endangered | Endangered | Vulnerable |
Criterion			
1. Undergone, suspected to have undergone or likely to undergo in the immediate future a reduction of numbers which is:	Very severe[a]	Severe	Substantial
2. Geographical distribution is precarious for the survival of the species and is:	Very restricted	Restricted	Limited
3. The estimated total number of mature individuals is: and either (a) or (b) is true:	Very low	Low	Limited
(a) evidence suggests that the number will continue to decline at a rate which is	Very high	High	Substantial
(b) the number is likely to decline and its geographic distribution is	Precarious for its survival		
4. The estimated total number of mature individuals is:	Extremely low	Very low	Low
5. Probability of extinction in the wild is:	50% in immediate future	20% in near future	10% in medium-term future

[a] all threshold levels defined in the guidelines to the Act.

numbers realistically is often problematical, and it is desirable to consider alternative grounds for conservation wherever possible and practicable. A further dilemma is that even the best estimates of insect population size are 'census' sizes, based on counts of all individuals present, and the relationship between these and effective population size is unclear. Generalisation is at present impossible but Frankham (1995), among others, considered that for invertebrates, effective population size may be as much as two orders of magnitude less than the census size. 'Effective population size', the number of individuals contributing to the next generation, is the more important consideration in conserving genetic variability within the population.

It is notable that, for the rather few Australian insects accepted for federal listing as 'critically endangered' or 'endangered', the majority were considered ineligible on the criteria of decline in numbers, population size, and probability of extinction in the wild, on the grounds that sufficient data were not available to demonstrate or quantify these characteristics. Most of these species were accepted for listing on the single criterion of 'precarious habitat' (criterion 2 of 5 in Table 1.2). Only for the Lord Howe Island stick insect (*Dryococelus australis*, p. 26) have these other criteria been sufficiently justified: with a field population 'unlikely to exceed 10 individuals', evidence for population decline and small population size is incontrovertible, but lack of quantitative data on probability of extinction precluded admittance against criterion 5. This approach essentially admits the problems of relying on quantitative data, and on inferences from unknown population dynamics, whilst not neglecting any such information available. Reliance on habitat features is much more tangible and, although debates may ensue about the adequacy of remaining habitat to sustain viable populations of the insect, allows an initial evaluation based on habitat characteristics, extent of loss, and threats.

Overcoming lack of population data: going beyond numbers

Even for the best known insects and regional faunas, exemplified by butterflies in western Europe, van Swaay and Warren (2003) essentially dismissed the quantitative IUCN criteria on population size and predictability of extinction on the grounds that they are impractical or irrelevant, with a clear statement of this belief. Thus, the criteria they used for establishing threat status differ in some important details from IUCN, as exemplified by the category 'critically endangered' (Table 1.3). They gave a valuable lead in giving priority to distributional changes, particularly to range contraction, as criteria revealed clearly by the extensive information on that fauna. In many parts of the world, the finer scale distribution of even the most popular insect groups is almost entirely unknown, and there is no historical background equivalent to that available in parts of

Table 1.3 *Criteria used by IUCN (1994) and van Swaay and Warren (1999) to determine the threat category 'critically endangered'*

IUCN	van Swaay and Warren
A. Population reduction of at least 80% over the past 10 years	A. Decrease in distribution of at least 80% over the past 25 years
B. Extent of occurrence less than 100 km^2 and two of the following: 1. Severely fragmented or known to exist at only a single location 2. Continuous decline 3. Extreme fluctuations	B. Present distribution less than 100 km^2 and two of the following: 1. Severely fragmented or known to exist at only a single location 2. Continuous decline 3. Extreme fluctuations
C. Population estimates less than 250 mature individuals and a strong decrease	C. For insects absolute numbers are rarely available and so less relevant
D. Population estimate less than 50 individuals	D. Not relevant
E. Probability of extinction at least 50% within 10 years	E. With the material available this criterion cannot be used

Europe. Evidence of such losses and declines is almost wholly from recent or obvious threats, sometimes augmented by anecdotal evidence. And, as noted earlier, some patterns of narrow range endemism may be more evident, as may be the loss of key biomes on which particular insects may depend – such as lowland native grasslands in southern Australia, and wetlands in many parts of the world.

Despite the usual alternative of erring on the side of caution in cases of doubt over an insect species' status, or deliberately incorporating parameters of uncertainty (such as 'fuzzy numbers') in quantitative status assessments, these steps may do little to enhance credibility.

One further caveat is that the act of uncritically listing butterflies or other 'collectable' taxa may draw undue attention to the species, leading to plundering by unscrupulous dealers (p. 130) for black market trading. A persistent difficulty is confounding 'rarity' with 'threat' or endangerment, with 'rarity' tending to attract attention and increasing commercial value of specimens. Terms such as 'rare', 'threatened' and 'endangered' have been used ambiguously and sometimes emotionally, making development of wholly objective criteria for assessing conservation status very difficult. It is perhaps worth reiterating that simple 'rarity', although often invoked in assessing conservation status and priority, is often irrelevant in practice, despite the popularly attractive connotations of the term. Numerous invertebrates are known from only very small areas, commonly from single sites and in apparently low numbers, but this condition (including connotations of narrow range or point endemism) is entirely normal, and does not necessarily indicate conservation need in the absence of perceived threats. Their circumstances may change rapidly, of course, due to stochastic or other events, but it is important to recognise that very many insect species may naturally have very small distributions. Various forms of rarity may indeed be associated with increased vulnerability, should threats arise. However, their absence from areas that appear to be suitable for them may not be due to anthropogenic changes but be entirely natural and reflect lack of some at present unknown critical resource, presence of a predator or parasitoid or other threat or, simply, that the insect has never reached the site. The skipper *Ocybadistes knightorum* (p. 78) in eastern New South Wales is one likely example of such narrow range endemism. Where it occurs, it is quite abundant (Sands 1999), but its range is limited by the narrow distribution of the sole larval food plant, the grass *Alexfloydia repens*. As another example, the New Zealand beetle *Prodontria lewisii* may never have occurred naturally beyond a distinct 500 ha region based on Cromwell sandy loam dunes, where it is now threatened by changes to vegetation and predation by vertebrates (Barratt 2007). It is now restricted to a single 81 ha reserve in central Otago.

The fundamental difference between the terms 'extent of occurrence' and 'area of occupancy' merits emphasis here, because the two are sometimes confounded, although clearly defined by IUCN and in some more local advices. 'Extent of occurrence' (Fig. 1.2a) is the area contained within the shortest continuous boundary that can be drawn to encompass all known, inferred or projected sites of present occurrence of a taxon, but excluding vagrants. It is represented commonly by the area of the minimum convex polygon which contains all the sites of occurrence. If there are major discontinuities in range – such as by large areas of obviously unsuitable habitat – these may be excluded. The more limited 'area of occupancy' (Fig. 1.2c) is the area within the extent of occurrence that is actually occupied by the taxon, and reflects that the species will often be distributed very patchily to reflect the uneven or patchy distribution of critical resources within the larger area. At one limit, the area of occupancy may be the smallest area needed to sustain a single viable population – and for some insects this can indeed be small, of a hectare or less. It can be measured in various ways, but is often evaluated through some form of grid mapping to separate 'occupied' from

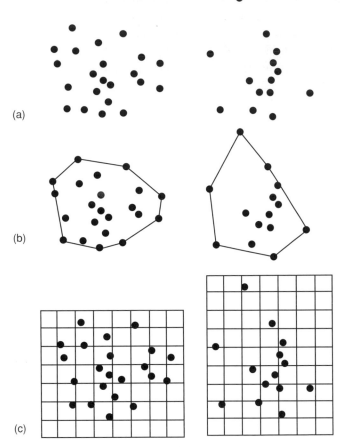

Fig. 1.2. Concepts of 'extent of occurrence' and 'area of occupancy' of a species, as used in IUCN (2001). Two examples are shown, in left and right columns, respectively: (a) the spatial distribution of known, inferred or projected sites of present occurrence; (b) one possible measure of the extent of occurrence, with the boundary enclosing all points in (a) above; (c) one measure of the area of occupancy, based on summing the grid squares occupied by the species to give a proportion of the overall extent of occurrence (after IUCN 2001).

'unoccupied' grid areas at some appropriate scale. Various size grids have been used, and periodic assessment can be an important monitoring (p. 200) tool. Thus, for the British butterflies, a 10 km × 10 km grid has long been the basic recording unit to assess major changes in distributions (Asher *et al.* 2001), with provision for subdivision of those relatively large areas (for example into 100 squares, each 1 km × 1 km) for finer detail.

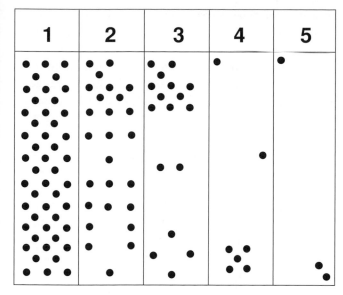

Fig. 1.3. Range composition of species: the distribution of colonies/populations of a species within the possible range (see text; after Kudrna 1986).

In practice the effective area of occupancy for an insect can sometimes be difficult to measure because, paradoxically, it can contain unoccupied habitat patches! The transient nature of included metapopulation segregates (p. 91), for example, means that currently unoccupied patches may indeed be part of the population's normal longer-term range and subject to recent extinction and future recolonisation. It is wise to infer at least the minimum set of areas that could be occupied by any species known to display a metapopulation structure. Formally, three features may be relevant in characterising a metapopulation (Thomas 1995), namely (1) occasional movements of insects between local populations, (2) colonisation and extinction of habitat patches fragmented within the landscape, and (3) local populations occurring in groups rather than single isolates. Such groups may be spread over at least several kilometres. These features are not usually confirmable without experimental investigation, but should be kept in mind during field appraisals.

Kudrna (1986) advanced the useful idea of 'range composition' for European butterflies. It is clearly applicable in many other contexts and at a variety of scales. This measure reflects the extent of occurrence and the degree of probable isolation and potential for interchange between populations. He defined five range models (Fig. 1.3), as follows.

1. Continuous or nearly continuous distribution over the species range.
2. Predominantly continuous distribution over most of the range with a small proportion of relatively isolated populations in some areas.
3. Predominantly isolated colonies with concentrated and more continuous distribution in significant central parts of the species range.
4. Discontinuous distribution over nearly the whole of the species range, so the species is largely in isolated colonies or populations, or with continuous distribution over a small part of the range.
5. Widely separated isolated single colonies or small groups of populations; generally with very restricted range within the possible arena.

Proposals to extend the range of a species or to increase the number of populations of many species in the latter categories are not uncommon, and intuitively wise in increasing their security. Many such species may indeed benefit from enhanced buffering or other protection of the sites on which they live, linked with prevention of any predictable impacts from threats that may arise, but the possibility of stochastic events (such as fire or storm) cannot constitute a universal basis for conservation management or action. This approach does not diminish the impact and importance of stochastic threats, but simply emphasises that they are difficult to predict and, therefore, to manage. Small sites are commonly associated with heightened vulnerability. The Schaus swallowtail butterfly (*Heraclides aristodemus ponceanus*) was listed as a threatened species in Florida in 1996, reflecting losses of its tropical hardwood hammock habitat and the influences of mosquito spraying programmes. It was upgraded to 'endangered' in 1984, following further decline. Two hurricanes (Hurricane Andrew in 1992, Hurricane Georges in 1998) severely damaged the remaining habitat, and butterfly numbers were reportedly reduced to an estimated few dozen. In this case, the effects were detected simply because the butterfly was already the subject of a continuing conservation programme involving monitoring of all occupied sites (p. 70). Existence and clarification of wider threat is universally more relevant, as Caughley (1994) emphasised some time ago, so that the causes of decline and their amelioration are the fundamental conservation need. In short, 'threat evaluation' gives us a basis for practical management and design of strategies to pursue this. The insect species that make their way onto threatened species schedules and the like on other grounds are much more difficult to evaluate. Most are indeed 'rare' (low abundance) and are inherently difficult to study, elusive, and often impossible to appraise comprehensively for impacts of likely threat. It is, perhaps, inevitable that confusion between rarity and threat will continue as a consequence of our inability to state confidently whether such an insect is indeed 'safe'. The 'Data deficient' category of IUCN (2001) has some valuable applications in such circumstances and it is here also that the ability to de-list species as new knowledge accumulates becomes invaluable (p. 229). In some legislations this step is difficult, whereas others provide for this to occur easily. Nevertheless, concerns exist that listing a species may be difficult to modify once it becomes law, and remain a permanent condition.

Occasionally, listings may seem contradictory within the same area. The same species can be assessed differently by different legislations, because of somewhat inconsistent criteria. Particularly when intra-country differences occur, the practical outcome of such vagaries can be a decidedly 'mixed message' that weakens the case for conservation, especially when those outcomes are based on the same core information. Australia's Mt Donna Buang Wingless Stonefly (*Riekoperla darlingtoni*) is one such example. It is listed under Victoria's state act as 'vulnerable' by being adjudged 'significantly prone to future threats which are likely to result in extinction' and 'very rare in terms of abundance and distribution'. This unusual stonefly is found only in a few small sites (at 1000–1200 m) near the summit of a single mountain in Victoria, and appears to be a genuine narrow-range endemic species there. The extent of occurrence spans no more than a few kilometres (Ahern *et al.* 2003) and the area of occupancy is estimated at around 2–4 km^2. Substantial targeted surveys of the area suggest that it is distributed very patchily, and an outlying population some 3 km from the main concentration appears to be extremely small. Mt Donna Buang has been added to Australia's Register of the National Estate on the basis of the stonefly's occurrence. The stonefly has been nominated for federal listing in Australia under the EPBC, but the advisory committee recommended that *Riekoperla* was not eligible for listing, on the major grounds of their not accepting that there are any known direct threats to the species, so that the habitat was not deemed 'precarious'. We thus have two different formal rulings on conservation status as outcomes for this species, based on essentially the same information, and consequent on different listing criteria. Adding to the complexity of definitively evaluating the species, Clarke and Spier (2003) indicated that 'it may be critically endangered'. *R. darlingtoni* has long been of conservation concern, and was one of the Australian insects included in the early IUCN Invertebrate Red Data Book (Wells *et al.* 1983), where it was categorised as 'rare'. In this, and similar cases, there is some danger that lack of a higher level (federal) listing may lessen priority for treatment at a lower (state) level of attention, in comparison with taxa that attain formal recognition at both levels.

'Declining populations' of insects (*sensu* Caughley 1994) are a more urgent concern than 'small populations' *per se*. Insects have been elected to 'threatened species' or 'protected species' schedules on two major grounds: a genuine need for conservation of the species, perceived on all available information, or a precautionary step for which evidence is less convincing or for which proponent zeal for listing has won the day. In both situations, outcomes have been rewarding. In some cases, listing has facilitated the additional work needed to reveal that the species is more secure than previously supposed or known and it can then be de-listed confidently to release resources for (then) more needy cases. Alternatively, focused conservation effort may lead to security, recovery, and decreased conservation status from threatened to non-threatened.

In addition, conservation becomes unnecessary once there is no reasonable doubt that the taxon has become extinct. However, as noted earlier, extinction is very difficult to prove. Zborowski and Edwards (2007) summarised the wider problem for much of the world in their assessment of Australian moths, when they noted (p. 26) 'No Australian moth is considered extinct but this is because so little is known about the distribution, ecology and distribution of moths. The fact that someone may have collected a species at Broken Hill in 1900 which has not been seen since does not mean that it is extinct but it does mean that no one who could recognise it has looked for it since in the right place at the right time'. This, of course, does not mean that it is necesarily *not* extinct! The Lord Howe Island stick insect (*Dryococelus australis*), for example, was for several decades believed to have been exterminated by rat predation, but was recently rediscovered on a small steep island, Balls Pyramid, near Lord Howe (Priddel *et al.* 2002), and has since become the focus of a major captive breeding programme (p. 182). As another example, the Otway stonefly, *Eusthenia nothofagi*, was listed as 'presumed extinct' in Victoria in 1991, following listing as Endangered by IUCN. However, after the discovery of a male in 1991, it has been found to be distributed widely in the Otway Ranges (Doeg & Reed 1995). Nevertheless, it is clearly endemic to that small area and it is thus important to conserve as a narrow range endemic species of potential wider value as a flagship species for wet forest communities in the region. Similar examples could be nominated for many countries and ecosystems, both terrestrial and freshwater.

Almost universally within this framework there is considerable lack of knowledge and, importantly, of understanding of the vagaries of invertebrate population dynamics, the significance of numerical fluctuations, pattens of apparency, and life histories. Many of the people responsible for promoting and undertaking conservation, or following up obligations flowing from the listing process, are expert in other aspects of conservation, or in very different groups of animals, such as larger vertebrates. The major issues for this aspect of invertebrate conservation thus centre around responsibility, both of listing (with the attendant difficulties of setting sensible priorities, almost invariably necessitating triage, and the neglect of other worthy species) on the best possible scientific grounds, and of the management actions that flow from that.

As discussed above, a major need in allocating a conservation status category to an insect is to recognise the impracticability of obtaining comprehensive population data; indeed, this is only rarely available even for extremely well studied abundant pest insects of massive economic significance to human welfare, let alone low abundance species. The complementary need is to avoid excessive detail, which can render the listing or categorisation process impossible or a severe deterrent to potential nominators. The requirement is then for alternative criteria that can be applied more easily, and by non-specialists, in ways that are both unambiguous and a sound guide to relative risk, as an aid to ranking species. Various attempts have been made to do this, or to suggest alternative categories for ranking, but most are underpinned firmly by the belief that the 'most needy' species should be accorded recognisable priority. For New Zealand Lepidoptera, Patrick and Dugdale (2000) outlined a set of criteria they considered useful for guidance in assessment, namely: uncommon or rarely encountered and with no historical evidence of declining populations or range decline; as above but with historical evidence of declines; known only from the type locality; uncommon or rare, biology unknown; type locality greatly altered; type locality at risk; host plant/site at risk, or predator influences seen in major parts of species range; genetic swamping of the endemic population by an adventive sister-taxon (relevant here in the case of an endemic lycaenid butterfly, *Zizina otis*, and an introduced close relative from Australia); no record of capture for more than 25 years, and now presumed extinct. Many of these criteria relate to rarity, and emphasise the importance of the type locality and population in a fauna with many narrow range species of uncertain taxonomic status. In this scheme, threats are indeed indicated but not given absolute priority. In a useful compromise, many botanists in Australia recognise a category termed 'ROTAP', an acronym for 'Rare Or Threatened Australian Plants', from the title of a pioneering report by Leigh *et al.* (1981).

Rarity and vulnerability

In that account, the many inconsistent ways in which the term 'rarity' can be applied are acknowledged, and the following qualification has many parallels in conditions for insects. Leigh *et al.* (1981, p. 11) wrote 'The decision as to when to treat a plant as rare has been somewhat subjective and made with consideration of a number of factors such as size and number of populations, size of total area over which the individuals are distributed . . . '. They noted the highly variable apparency of some species and commented, further 'Such species are not listed as rare . . . if they are known to be quite common at times', and concluded the qualification of rarity by noting that population trends for seasonally fluctuating species can be difficult to interpret without special long term studies and that some species in this class may indeed be under threat, with the impact of processes such as grazing not yet recognised. Later (Leigh *et al.* 1984), 'rare' was emphasised as applying to species of plants not currently considered endangered or vulnerable, but represented by a relatively large population over a very restricted area or by smaller populations spread over a wider range, or some intermediate combination of these patterns. The first of these conditions is often referred to by entomologists as 'local', but there are many clear parallels with insects in the above perspective.

Rarity can clearly predispose a species to threat, and be associated with increased vulnerability, for example, by localised changes to habitats that would not affect a widely distributed species. Even though the term may not be defined precisely in many individual conservation contexts (and may not equate with threat) it seems destined to remain an important focus for conservation activity, as raising both emotive and practical issues. In practice, 'rarity' of an insect is a condition that is often prejudged on the basis of the species being known from few individuals or localities but (following Rabinowitz *et al.* 1986) with the three 'axes' of low abundance, small distribution and ecological specialisation. As Abbott *et al.* (2007) noted for the endemic Western Australian earwig fly (*Austromerope poultoni*, Mecoptera), before judging an insect as rare it is prudent to search thoroughly using appropriate detection or sampling techniques. *A. poultoni* was not collected for 60 years after its initial discovery. It is cryptic and lives in leaf litter, and Abbott *et al.* suggested that, possibly, unsuitable methods were used in attempts to find it subsequently or it was simply regarded as too rare to find again. Their recent surveys revealed this scorpionfly to be quite broadly distributed in the southwest of the state. Specimens were found at 15 of 48 localities investigated, and across a variety of forest types; most individuals were taken in pitfall traps. It is common in entomological studies for additional or novel collecting methods to yield species regarded previously as very rare or elusive – so that 'rarity' is simply an artefact arising from our ignorance over its biology. However, many insects *are* indeed rare and their status simply endorsed by persistent failure to discover them during additional targeted surveys.

There is some advantage in having only a few categories of potential conservation status to consider, rather than a longer series of categories between which boundaries become blurred, and ambiguities can proliferate. The United States Endangered Species Act recognises only two categories for species status allocation: 'endangered' ('in danger of extinction throughout all or a significant proportion of its range') or 'threatened' ('likely to become an endangered species within the foreseeable future throughout all or a significant part of its range'). Secondary lists, of 'species of concern', 'species at risk', 'sensitive species' and others raised in many parts of the world, tend to reflect taxa for which concerns have been raised but for which biological information is lacking. Categories may be absolute (each of the IUCN categories) or relative (when used for ranking). Although the IUCN categories constitute a valuable hierarchy of relative conservation need, their call for incorporating quantitative thresholds is a distinct deterrent for their adoption for insects and other invertebrates. The approach suggested by Sands and New (2002) for Australian butterflies is a step towards overcoming this, but includes provision for considering extinction risk against a series of time intervals and qualified by a series of conditions, not all of which need to be filled. Priority was given to detecting threats, as a basis for instituting constructive management, and also to determining whether populations occur in currently protected areas (such as National Parks) where site protection could be assured as a major requirement for management. Their categories were expressed within a key, as follows.

1. Information on biology, distribution and resident/vagrant status sufficient to make an informed evaluation of conservation status . . . 2
   – Information insufficient to make an informed evaluation: with little or no information on any of the above topics . . . Data Deficient (DD)

2. Threats defined for the species; threats to major populations or population segregates likely to lead to species extinction . . . 3
   – No threats defined for the species: threats to major populations or population segregates not likely to lead to species extinction . . . No Conservation Significance (NCS)
3. Threats identified for all known populations and considered to pose a risk of extinction within 5 years (one or more listed conditions implicit), usually no more than 5 populations or major population segregates known . . . Critically Endangered (CR)
   – Threats identified for all or most known populations, normally including those of greatest significance (size, distribution), and considered to pose a risk of extinction within 5–10 years (one or more listed conditions implicit) . . . Endangered (EN)
   – Threats identified for some populations and considered sufficiently important to pose a risk of species extinction within 10–20 years (one or more listed conditions implicit) . . . Vulnerable (V)
   – Threats identified but not considered to pose a risk of species extinction within 20 years (one or more listed conditions implicit) . . . Lower Risk (LR)

The 'listed conditions', summarised in Table 1.4, allow for inclusion of both absolute information and informed supposition, with the aim of the criteria being practicable rather than subject to continuing debate based on quantitative thresholds.

Listing criteria designed predominantly for vertebrates and other organisms are sometimes difficult to apply confidently to insects, following from the above lacunae in knowledge, and with the realisation that it is virtually impossible to extrapolate any available quantitative information on the population dynamics of one insect species to any other, even if closely related, or, even, between conspecific populations in different sites. Some of these problems of determining conservation status merit illustration here.

First, many insects are available as adults for survey for only short periods, perhaps a few weeks, each year, with much of their life passed as cryptic or inaccessible or unrecognisable immature stages. Second, even when adults are present, sound quantitative data on numbers can be difficult to obtain, and even presence/absence may be problematical to determine. As an example, consider the golden sun-moth, *Synemon plana* (Castniidae), an important flagship insect for native grasslands in mainland southeastern Australia, as a species that poses a number of

Table 1.4 *Criteria for categories of threat designated for Australian butterflies*
See text for details.

Category	Criteria/listed conditions
Extinct (EX)	Extinction of all historically known populations of a species. A species which has not been found in any documented habitat where it was formerly present, or elsewhere, despite targeted searches and surveys over an extended period (*c.* 20 years)
Critically Endangered (CR)	(a) no populations known in protected areas (b) species known from one or very few populations (c) evidence of population demise or loss of breeding sites (d) evidence of decline in area of occupancy, and none of expansion Recovery measures deemed urgent
Endangered (EN)	(a) threats less severe than CR, and in combination with: (b) none, one or few breeding populations in protected areas (c) management inadequate for reducing threats of extinction
Vulnerable (V)	(a) threats not necessarily as severe as for EN, commonly varied across different populations, and in combination with: (b) none, one or few breeding populations in protected areas (c) small number of populations and/or small range of occurrence, with or without evidence of decline in area of occupancy (d) management inadequate for reducing risks of extinction
Lower Risk (LR)	Not categorised as at risk of extinction (a) some threats recognised but not well defined for all populations (b) usually localised or limited range species, ecological specialists and (c) signalled as of conservation interest because of documented decline in area of occupancy or range, or failure to discover additional populations through targeted surveys (d) some populations may be in protected areas
Data Deficient (DD)	Knowledge of biology/ecology insufficient to allow assessment of conservation status: reasons may include being known from very few individuals, ambiguous province or label data, no evidence of breeding populations, lack of targeted survey over parts of potential range

Source: Sands and New (2002).

biological features that render sampling difficult or misleading. On any given site, the flight season of this moth extends over about 6–8 weeks, the precise period depending on latitude and weather conditions. At other times of the year, the caterpillars are underground. They feed on grass roots and are completely inaccessible for enumeration without digging them up. Most sites on which S. *plana* occurs in Victoria are small (a few hectares, or less), but some clarification has come from a survey of the moth on the largest site on which it is known to occur in Victoria (Craigieburn Grasslands, north of Melbourne), using a series of spot counts throughout the flight season and a belt transect approach involving 24 people for a single more extensive survey (Gibson & New 2007) over a grid of 120 m × 1200 m. Consider the following: (1) individual moths cannot feed, and live for only 3–5 days, so that any single occasion sample, however accurate it may be, reveals only a small cohort of the entire population on the site; (2) the rapid turnover renders any form of mark–release–recapture exercise of very limited value; (3) only male moths are sufficiently active for detection, and fly only in short bursts and for short distances (normally a few tens of metres, at most); (4) they fly only between about 1100–1400 h, so that surveys at other times may not reveal them and only under suitable weather conditions – if cloudy, rainy or windy, or at air temperatures lower than about 20 °C they are inactive; (5) over the flight season, local 'hot spots' of abundance on a site occur at different times, probably relating to local microclimate differences such as insolation affecting local emergences; and finally (6) it is not yet confirmed whether the moth is univoltine or has a life span of 2 or even 3 years – in which case even a wholly accurate single season survey may account for only one (unknown proportion) cohort of a resident population on a site.

In a somewhat different context, some insect species are highly irregular in occurrence across generations, so that abundance/incidence trends are extremely difficult to interpret even between successive generations. One such species is the Australian fritillary butterfly (*Argynnis hyperbius inconstans*), which aroused considerable debate in attempting to assess its conservation status accurately for Australia's Butterfly Action Plan (Sands & New 2002). Different people experienced with the fritillary in the field were variously adamant that it was either critically endangered or of little conservation concern in parts of its limited coastal range, and part of this dilemma devolves on the butterfly's highly variable apparency. It appears to undergo 'boom and bust cycles' in which sequences of seasons in which it is extremely scarce (or apparently absent) on sites are punctuated intermittently by seasons in which considerable numbers occur. It is possible that an irregular diapause facilitates a process of 'population accumulation', whereby in 'poor' seasons many larvae do not progress to adulthood, but numbers accumulate over several generations to result in a 'mass emergence' in more favourable conditions, such as those triggered by rain. Irregular diapause regimes may be quite widespread in insects, but are difficult to interpret or detect (Sands & New 2008).

Many insects are truly univoltine, and such differences are thus superimposed on a largely predictable seasonal pattern of development and apparency. Other species can normally take longer to develop: the saproxylic larvae of the British stag beetle, *Lucanus cervus* (below) take up to six years to reach maturity. In short, often 'what we see' in insect surveys at any time or within any single season of sampling is not a valid representation of 'what is there'.

Inference from these examples, which could be multiplied extensively and expanded to include other interpretative difficulties and ambiguities, is that establishing the precise conservation 'status' of many insect species is remarkably difficult, and that any dependence for this on even reasonable data on population size and variability is often not realistic in planning conservation. Even obtaining basic evidence to determine their presence or absence, such as simply by seeing an individual of a rare species, may prove difficult. New and Britton (1997) reported sighting only five individuals of the lycaenid *Acrodipsas myrmecophila* during three seasons of field work at the only site at which it was known to occur in Victoria, and noted that 'determining the presence of the butterfly each season is therefore costly and uncertain'. For such situations, obtaining any quantitative population data is clearly impracticable. For background, consider also that for even the most intensively studied of all insects in Australia, some of the major agricultural pests such as *Helicoverpa* moths, detailed data on population dynamics from which to forecast abundance and on which to found predictive models on population changes to formulate suitable pest managment needs are often also highly uncertain.

As another example, a 10-year monitoring programme for Endangered, Vulnerable and Near-threatened saproxylic beetles in Finland (Martikainen & Kaila 2004) revealed most such species only very patchily. Two endangered species were represented only in three years and one year, respectively, and on each occasion by a single individual. No 'Vulnerable' species was trapped in more than six of the ten years.

Some insects are, of course, far more predictable and constant in appearance than *Synemon* or *Argynnis*, but it is very common for numbers in a given insect population to naturally fluctuate severalfold over successive generations, so that even long-term (multiseason) trend analyses may not be particularly informative without considerable biological understanding and insight.

Box 1.3 · *Functional conservation units in a butterfly:* Maculinea alcon *in Belgium*

Maculinea alcon (Alcon blue) is a highly sedentary lycaenid relying on a single larval food plant (marsh gentian, *Gentiana pneumonanthe*) but associating with different species of *Myrmica* ants in different parts of its European range. A recent study in Belgium (Maes *et al.* 2004) sought to determine 'functional conservation units' (FCUs) for the Alcon blue, at different spatial scales. Detailed distributional data on the butterfly, food plant and habitat, including population size (based on counts of eggs) and estimates of mobility and colonisation capability, were used to define FCUs on three scales: (1) the 12 presently occupied habitat patches plus the area within 500 m range surrounding them; (2) the areas within a range of 2 km around the occupied patches; and (3) potential recolonisation sites on which *M. alcon* has recently become extinct. The first two of these were defined in terms of (1) the maximum local movement distance based on mark–release–recapture information and (2) the maximum observed colonisation distance.

FCUs were a valuable tool in ranking species conservation measures for priority, with the needs for wet heathland areas and high numbers of *Gentiana* recognised, so that potential restoration sites were categorised further as currently suitable or potentially suitable after restoration. Conservation objectives were defined for each FCU category:

1. FCU1. The main objective for these is to increase butterfly population size by optimising habitat conditions, by enlarging existing habitat patches and restoring potential patches. A combination of conventional maintenance management (such as by low-intensity grazing and small scale burning) and intensive care is needed to increase density of *Gentiana* and of *Myrmica* nests, using techniques such as very small-scale (1 m^2) sod-cutting. Excluding grazing stock

from 15 July to 30 September from *G. pneumonanthe* patches is a significant measure to protect *M. alcon* eggs.

2. FCU2. This category is based on heathland patches within 2 km of FCU1s being likely to be colonised naturally, so that restoration measures are based on 'stepping stones' to create new habitat, and increasing connectivity is necessary.

3. FCU3. These are areas potentially habitable by *M. alcon*, as candidates for re-introduction, and are either suitable at present or in need of restoration to render them so. Maes *et al.* pointed out that, in the absence of the butterfly, management can be intensive – simply because harm to an existing population is not a consideration. Nevertheless, surveys for suitable ants are 'highly relevant' before starting this, as these may need to be conserved through mosaic management. The optimal host ants in Belgium are *Myrmica ruginodis* and *M. scabrinodis*.

Maculinea alcon was the first invertebrate species in Flanders for which an action plan was prepared and, in conjunction with the Maes *et al.* (2004) study, incorporates the dual aims of enhancing the viability of existing populations and creating new ones on additional sites. Progress depends on continuing logistic support, and part of the rationale of designating FCUs was to consider the different spatial scales and priorities to enable constructive progress to be made.

Species and related conservation units

Even recognising our focal species/taxon may be difficult because of taxonomic inadequacies or ambiguities caused by patterns of variation and, not uncommonly, by the lack of availability of specialists able to define or clarify these entities. This recognition is central to listing, where the taxon must be recognisable and diagnosable as a 'real entity'. However, no absolute definition of 'a species' in any of the various ways in which the concept can be defined is usually specified, so that even the apparently simple task of defining 'a species' may be problematical. In many legislations, that entity must carry a scientific name; in others, undescribed species can be recognised formally (as with the ant '*Myrmecia* sp. 17' in Victoria) as long as voucher specimens for reference are responsibly deposited in an institutional collection and so available for future reference. Many of the taxa referred to in this account, particularly of butterflies are 'subspecies', sometimes the subjects of continuing debate over their precise status. Occasionally, accusations arise over 'inflating' the status of such taxa in conservation by not recognising possible synonymy (and so downgrading their significance by including them as members of more widely distributed and secure taxa) with the aim of assuring their eligibility for funding and other support. Even the correct status of putative full species, such as the Brenton blue (*Orachrysops niobe*, p. 207) may be controversial, with specialists divided over the specific integrity of many such entities. In many legislations, 'subspecies' are deemed fully equivalent to full species, provided that there is consensus among experts over this stance, and the characteristics that diagnose that entity are at least reasonably unambiguous. The practical need is for any undescribed taxon listed to have been studied sufficiently that it can be 'clearly and reliably distinguished from other known taxa' (Mawson & Majer 1999).

Box 1.4 · *Species or distinctive populations? Some examples from North American tiger beetles of conservation importance*

Pearson *et al.* (2006) estimated that at least 15% (33 of 223) named species or subspecies of tiger beetles (Cicindelidae or Carabidae: Cicindelinae) may be declining at rates sufficient to merit consideration for listing under ESA. However, only five were then officially listed, with seven others under consideration. Many of these beetles are very localised, some known from single sites, and the distinctiveness of several taxa has been an important component of determining their conservation status. Genetic studies have been a vital tool in this endeavour and in determining relationships between taxa and populations. Collectively, these cases have helped to understand ways in which such insect populations may be characterised and their evolutionary significance appraised. The following three cases exemplify the significance of such studies.

1. The Puritan tiger beetle (*Cicindela puritana*) is known in New England from only two populations (Connecticut River, Chesapeake Bay). These are strongly distinct on genetic grounds, to the extent that Vogler *et al.* (1993) recommended that they should be subject to different management regimes.
2. The entire known population of the Coral Pink Sand Dunes tiger beetle (*Cicindela albissima*) occurs within a 400 ha site in southern Utah, where its population size fluctuates within the range of about

800–2000 individuals. The site is threatened by recreational vehicle use, a widespread threat for habitats on which many tiger beetles depend. *C. albissima* was formerly considered to be a subspecies of the more widespread *C. limbata*, of which the other four subspecies are more widespread in western North America. *C. albissima* is the only one of these found west of the continental divide. A study of this group using mitochondrial DNA (Morgan *et al.* 2000) showed the distinctiveness of *C. albissima*, and led to its resurrection as a full species long isolated from other members of the complex.

3. A rather different approach was used by Goldstein and De Salle (2003) to characterise populations of the Northeastern Beach tiger beetle (*Cicindela dorsalis dorsalis*). Single hind legs from museum specimens were used for DNA extractions for polymerase chain reaction study. Widespread polymorphism was found from the 42 specimens yielding useful results, prompting questions about the most suitable stock to use for re-introducing the beetle in New England. The main question was whether to use existing populations (from Massachusetts) in order to maintain local uniqueness or to employ more southerly populations, which would facilitate restoration of past genetic diversity. In this case, Goldstein and De Salle noted that arguments could be made for either option but, on ecological characteristics (such as adaptations to ocean-front storm regimes) opted for the former approach (see Box 7.4, p. 186).

More general early background to the context and approaches of assessing evolutionarily significant units by using tiger beetles is given by Vogler and De Salle (1994).

Reflecting the commonly observed patterns of intraspecies variation and narrow distributions, the concept of 'significant populations' has particular value and relevance in insect conservation. Again, this term is usually not defined formally, but has connotations of populations of particular scientific or strategic importance, and may include populations that are key source populations for breeding or dispersal, populations necesssary to maintain genetic diversity and populations near the edges of a geographical range. They are thus usually of species that are not wholly endangered but for which those populations are threatened. Such populations may, for example, be geographical outliers from the species' predominant range and so especially isolated. They might, perhaps, represent a particular phenotype or host plant/resource specificity, be the extremes of a clinal pattern, exhibit unusual genetic features such as those likely to lead to its recognition of a distinct taxon in the future or to key evolutionary understanding, and so on. For any such case, the justification for conservation is needed on scientific grounds acceptable by broad consensus. Significant populations may or may not be regarded as taxonomically distinct. The Australian skipper butterfly *Hesperilla flavescens flavescens* has this subspecific name applied to a few populations limited to saltmarsh/sedgeland habitats west of Melbourne and for which the butterfly is very distinctive in appearance. However, it is probably one extreme of a pattern of clinal variation with a complex involving *H. donnysa*, itself very variable. As with the South Australian *H. f. flavia*, caterpillars of *H. f. flavescens* feed on the sedge *Gahnia filum*. Crosby (1990) investigated the pattern of distribution of this complex and confirmed that the populations near Melbourne represent 'the extreme yellow phenotype'. Those populations were designated important as 'reference sites' for studying the evolution of the complex, and Crosby recommended that they be regarded (by analogy with Britain) as equivalent to sites of special scientific interest (sic). Uncertainty over the taxonomic position of this skipper initially delayed it from being listed for protection in Victoria.

The general term 'evolutionarily significant unit' has sometimes been used to infer or designate a population of fundamental importance to understanding or conserving a taxon, as assessed from the best available scientific evidence. Two broad criteria are important in deciding whether a population merits this formal equivalence to a 'proper species', and these must be justified under some legislations. These are (1) that the population is reproductively isolated from other populations of the same taxon, and (2) more subjectively, that the population must be an important component of the species' evolutionary legacy. Opinions on the latter are likely to differ widely, but devolve on the issue of 'distinctiveness', be this distributional, genetic or phenotypic. Precise formal definition is usually impracticable. Confusion occasionally arises through populations that are simply 'political outliers' in regions subject to multiple jurisdictional attentions. Some Australian or European butterflies, for example, have ranges that extend narrowly across State borders or country borders, respectively, so that they are abundant in one State/Territory or country but extremely scarce in the neighbouring one. The jurisdiction of the latter may necessarily have to consider its status only within that area, and not its wider level security. Common sense must then prevail over any more fundamental measures for conservation priority. The dilemma also arises across many levels of jurisdictional application, when species extend marginally from any politically defined area in which they are common, to others. They may be signalled as of national conservation significance at the edge of their range, particularly when conditions there are only marginal for them to thrive. In Europe, the nymphalid butterfly *Euphydryas desfontainii* was noted some years ago (Descimon & Napolitano 1992) as represented in France by a single endangered colony, but to be common in Spain (and North Africa). Descimon and Napolitano queried the value of strenuous efforts to conserve the French population within the wider perspective of European priority needs. The associated concept of 'Range Affinity' advanced by Kudrna (1986) for European butterflies bears on this, particularly the levels of 'Quasi-European' and 'Quasi Extra-European' used to respectively include taxa extending narrowly into Europe from elsewhere and narrowly elsewhere from Europe.

Inferring and defining threat

Parallel dilemmas occur once we move on to look at management aspects arising from designation of a species as 'threatened' in some way. However, it is extraordinarily easy to overlook some important threat or component of threat, and any conservation plan must include the fullest possible consideration of these. For illustration, Yen and Butcher (1997) listed 13 primary threatening process categories for non-marine invertebrates (Table 1.5), with many of these processes overlapping and interacting in various ways. These categories are all broad, but compiling a 'checklist' of possible threats and assessing their relative importance and impacts for any given insect species is a valuable exercise (see p. 216). Not least, this exercise ensures (as far as possible) that no significant threat has been overlooked at this important stage. Each can then be analysed in greater detail, and at appropriate scales, to determine more specific needs for abatement and management. Thus, for the Bathurst copper butterfly (p. 217) a stated criterion for the published recovery plan is that 'Factors detrimentally affecting the Bathurst copper butterfly or its habitat are known for each of the sites within two years'. These factors are listed collectively as 'loss of habitats, illegal collection, firewood collection, feral animal activity, and fire management practices'. As in other contexts, threats may be divided functionally into those that kill the insects directly (pesticides, overcollecting: p. 129) and the far wider array that affect the habitat and resources needed by the insect and so

Table 1.5 *Categories of threat to Australian non-marine invertebrates and some examples*

Category[a]	Notes and examples
Agriculture and clearing of native vegetation	Large scale: soil erosion, salinity, exotic species, chemical inputs (etc.)
Habitat fragmentation	Landscape level effects, isolation, genetic impoverishment
Grazing and trampling	Loss of native vegetation, changes to water bodies
Inappropriate fire regimes	Frequency, intensity, scale, seasonality (etc.)
Forestry activities	Variety of effects; exotic species, loss of diversity, changed hydrology, changed soil quality (etc.)
Pollution	From industry, agriculture, urbanisation; increased nutrient inputs, toxins, temperature changes to water
Exotic and adventive taxa	Invasive and feral plants and animals
Alterations to aquatic ecosystems	Riparian vegetation, regulation of flow, draining of wetlands, pollution, exotic taxa (etc.)
Mineral extraction	Site-specific effects, habitat changes
Transport and recreation	Road construction: fragmentation, loss of connectivity; resort development and access issues: coastal development; winter sports development in alpine areas
Pests and disease	Introduction and facilitation of spread; non-target effects
Direct exploitation	Possible overcollecting, bycatch issues
Long-term environmental change	Climate change: future distribution and accessibility of resources

[a] *Source:* Yen and Butcher (1997).

decrease the quality or extent of its environment. Fully objective criteria for selecting insect species for priority treatment on threat evaluation are commonly lacking, and there may be great benefit in using the precautionary principle judiciously in cases where such detail is unavailable. This principle helps to safeguard species over which we remain ambivalent or ignorant. Some cases of endangerment are very clear, but many contain substantial inference. Any subsequent actions must help to focus attention more precisely. At present, the conservation needs of most insects rely on qualitative judgment of experts, and capability to assess this reliably varies enormously across different insect groups and regions. Informed consensus is often difficult to obtain.

The term 'significant threat', used above, is difficult to define but it is often necessary to decide as objectively as possible whether a suspected

Table 1.6 *Significant impact criteria for evaluating threats to species*

An action is likely to have a significant impact (and, so, be a threat) if there is a real chance or probability that it will cause one or more of the following outcomes.

Lead to a long-term decrease in the size of a population (CR, E) or important population (V).

Reduce the area of occupancy of a species (CR, E) or important population (V).

Fragment an existing population (CR, E) or important population (V) into two or more populations.

Adversely affect habitat critical to the survival of the species (CR, E, V).

Disrupt the breeding cycle of a population (CR, E) or important population (V).

Modify, destroy, remove, isolate or decrease the availability or quality of habitat to the extent that the species is likely to decline (CR, E, V).

Result in invasive species that are harmful through becoming established in the species' habitat (CR, E, V).

Introduce disease that may cause the species to decline (CR, E, V).

Interfere (CR, E) or interfere substantially (V) with the recovery of a species.

Interfere with a reintroduction into the wild (EW)

Adversely affect a captive or propagated population or one recently introduced/re-introduced to the wild (EW).

Source: (after DEH 2006; IUCN categories are given as EW (Extinct in the Wild), CR (Critically Endangered), E (Endangered), V (Vulnerable)).

threat will indeed have a 'significant impact' on the species. Guidelines for this are sometimes available (Table 1.6), encompassing a variety of criteria linked to threat category with the general principle that 'the general test for significance is whether an impact is "important, notable or of consequence, having regard to its context and intensity"' (DEH 2006).

A serious and recurring practical problem is that for many insects decline has been detected or inferred but the precise reasons for this remain speculative rather than being attributable directly and unambiguously to a specific threat or suite of threats whose mitigation can then form the basis for informed management. If threats cannot be defined to this extent, management is almost inevitably less focused. The North American burying beetle *Nicrophorus americanus*, for example, has declined dramatically over around 90% of its former range (Fig. 1.4), and Sikes and Raithel (2002) attempted to determine the reasons for this by reviewing the eight major hypotheses of cause of decline put forward by that time (Table 1.7). These encompassed pesticide use, attraction to artificial lights, impact of pathogens, and a suite of themes related to habitat loss: whether the beetle is an old growth forest or prairie specialist (so that

Table 1.7 *Hypotheses advanced for the decline of the American burying beetle*, Nicrophorus americanus

References for each hypothesis are included in Sikes and Raithell (2002); 'accounts for congeners?', does the hypothesis explain why sympatric congeners are not affected?; 'accounts for pattern?', does the hypothesis explain the geographic pattern? (Fig. 1.4).

Hypothesis	Accounts for congeners?	Accounts for pattern?
DDT/pesticide use	No	No
Artificial lighting	No	No
Pathogen	Yes	Yes
Habitat alterations:		
Old growth specialist	Yes	No
Prairie specialist	Yes	Yes
Vertebrate competition	Yes	No
Loss of ideal carrion	Yes	No
Congener competition	Yes	No

Source: after Sikes and Raithel (2002).

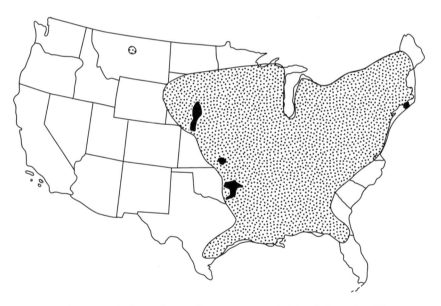

Fig. 1.4. The range decline of *Nicrophorus americanus* in North America. The stippled area indicates historical distribution (extent of occurrence); the solid black areas indicate recent/current distribution (after Sikes & Raithel 2002, with kind permission of Springer Science and Business Media).

loss of either might be deleterious), habitat fragmentation affecting the supply of vertebrate carrion needed by the beetle, or changing the quality (species composition) of that resource. The last two suggested likelihood of increased competition from congeneric beetle species. Even the best supported of these ideas required further studies for clarification of why *N. americanus* had declined, so that the success of conservation efforts cannot at present be predicted confidently. In cases such as *N. americanus*, reviewing all possible causes of decline objectively is a useful exercise; in this case, background considerations included whether congeneric species were also affected, and whether each hypothesis is supported by changes in geographical range. Some hypotheses will remain difficult to evaluate – many of the extant populations of *N. americanus* occur in remote, lightless areas and it is difficult to entirely dismiss 'artificial lighting' as a contributor to past mortality and decline closer to human settlements.

The above examples illustrate a problem that goes to the very heart of credibility in insect species conservation: that simply gaining sufficient biological understanding of low abundance species as a foundation for sound evaluation of threats and for designing management can be extraordinarily difficult. It is far easier to study insects that are abundant, and for which manipulative field experiments can be undertaken without fear of exterminating them. Many studies on insects of conservation concern must be based on observation (with due regard to not unduly damaging the habitat – by excessive trampling, on small sites, for example) and correlations, because of the substantial risks likely to occur from more interfering or manipulative experiments. Likewise, truly replicated and statistically convincing observations or tests may be impossible on single small sites or with small populations. This reality must be accepted, but is a source of frustration when we realise that the best possible scientific basis for our conservation actions may not be founded in the ideal scientific endeavour, but in compromise. This limitation also means that the general ecological background and lessons from other, more common and accessible, insects related to our target species, or from parallel studies on other taxa, can provide very relevant background to help interpret information on the species we seek to conserve.

Further focus and need

Most insect species for which some reasoned concern arises or for which conservation management is contemplated are thus those that have demonstrably declined in abundance and/or distribution, or that are believed likely to do so imminently as a result of impending threats. However, the aims of management may range from simple protection of 'what is left' (with the emphasis on removal of threats, prevention of their reccurrence, and prevention of any further losses) to more aggressive 'recovery', with the aim of increasing numbers of individuals and/or populations and/or extending distribution. The first of these aims is clearly implicit in the second, as the basis for recovery, and is founded commonly in some form of protecting occupied sites from further disturbance and loss of area or resources. The second approach seeks to extend or enhance these in some way, and also encompasses exercises such as translocation or re-introduction (p. 168), with attendant considerations over how and where to pursue these measures.

The initial planning decision, driving the scope of any management plan, is thus simply to elect for one or other of these approaches. An important caveat is that, unless threats to a species or population deemed as 'protectable' rather than needing recovery are detected, the target species may continue to decline until the second option becomes paramount. Monitoring is therefore essential, and should be directed at the supply of critical resources as well as the numbers and distribution of the insect involved. As we note later, the dynamics of the insect's host plant or prey, mutualistic species and natural enemies are all components of the effective environment that must be sustained. The level of support needed for conservation must also be assessed realistically at an early stage; many worthy conservation plans for insects have withered for lack of continuing commitment.

Public interest and support is a key element in helping to sustain commitment to species conservation. Simply that a species is liked or adopted as a local emblem and can be promoted as a local 'flagship' can assure local interests in its continued wellbeing. In Japan, the nymphalid butterfly known as the 'Great purple emperor' (*Sasakia charonda*, sometimes regarded as Japan's national butterfly) has become a symbol for conservation in several places. In Saitama Prefecture, one school built a large observatory to rear the species (Makibayashi 1996), and it is the recognised symbol of the town of Nagasaka-cho in Yamanashi Prefecture (Bandai 1996). Several similar examples of regional adoption are noted later.

'Liking' an insect can thus support conservation interest but is not on its own a politically or ecologically persuasive ground for according priority. The major grounds for the latter are based most commonly on extent of threat, as emphasised above, with selection of particular insect groups favoured by better biological knowledge and understanding. A rather different perspective, advocated by Haslett (1997) in the context of Europe's Bern Convention, is that some insects may be selected for priority on the grounds of representing or frequenting habitats that are under-represented among previously listed taxa. Haslett (1997) emphasised that there are far too few invertebrates listed to provide a reasoned perspective of the group's conservation needs, and suggested that additional focus was wiser than simple more random proliferation with emphasis on still more taxa from the same more familiar suite of habitats – without in any way diminishing the importance of the latter approach. He suggested that habitats such as caves, running waters and dead wood all support numerous insect species susceptible to change, and that more attention to those species would help to draw attention to the taxa that are necessary to ensure the continuity of those commonly overlooked ecosystems.

The great majority of insect species signalled for conservation protection are herbivores, and species at higher trophic levels, as predators and parasitoids, are even more greatly under-represented. The only major exception is of some of the myrmecophilous lycaenid butterflies, some of which feed as predatory caterpillars on ant brood. The wellbeing of predators and parasitoids will commonly depend absolutely on that of their insect prey or hosts, which may be very specific. Shaw and Hochberg (2001) reiterated a comment by Shaw (in Shirt 1987) that parasitoid Hymenoptera must be considered among the most threatened of British insects, but also that even attempting to make any list of priority or endangered species is 'hopeless' because of poor knowledge. Shaw and Hochberg (2001) suggested that insects at these higher trophic levels may be both intrinsically (through the fate of their resource species) and extrinsically (through lack of regard and attention to their status) extinction-prone. They suggested that parasitoid Hymenoptera exhibit a number of features that render them particularly prone to local extinctions. In many species of these wasps, inbreeding can lead to male-biased populations at low densities, many taxa have a very high level of host specificity, and adult behaviour can be influenced strongly by climatic conditions – so that small populations may be lost simply through experiencing bad weather conditions at certain times. Ethically, a species of parasitoid wasp or fly is an entirely valid target for conservation – just as much so as a more popular and appealing butterfly or dragonfly. Yet, in seeking to conserve the 'eaten' it may happen that the 'eaters' are themselves considered a threat, and controlling or reducing their impact may be viewed as a key management component to pursue. Ectoparasites such as fleas or lice do not generally cause mortality, and their conservation devolves entirely on conserving the vertebrate hosts on which they depend. The poor image of 'parasites' is enhanced by the general perception that they are 'bad' and must be eliminated (Windsor 1995). However, sympathy for many insect predators and parasitoids is increasing through their perceived importance in 'conservation biological control', through which native species are valued progressively in the control of numerous agricultural pest arthropods (Hochberg 2000; New 2005a). Nevertheless, this perception is based on the role of the wider feeding guilds, rather than individual threatened species, and very little interest in this facet of parasitoid conservation has emerged, despite widespread realisation that probably all threatened species of Lepidoptera and other phytophagous groups have specialised parasitoids that will suffer 'co-extinction' (Stork & Lyal 1993) should their host be lost.

The uneven attention given to parasitoids is exemplified by Hochberg's (2000) example that the lycaenid butterfly *Maculinea rebeli* is a formally recognised (i.e. 'listed') threatened species throughout its European range (p. 59), whereas its specific parasitoid ichneumon wasp (*Ichneumon eumerus*) is not acknowledged in this way. This wasp is believed to be a complete specialist on this single host species. It is thereby likely to be at least as endangered as *M. rebeli* and, because it does not occur with all host populations (Hochberg *et al.* 1996), in reality even more so. The significance of the parasitoids of *Maculinea* includes that they were among the first such insects noted specifically as of conservation interest. Thomas and Elmes (1992) noted that the parasitoids generally occurred on only a minority of *Maculinea* sites and were seldom common, even there.

The factors that determine the patchy incidence of the wasps are not yet understood. There is no reason to suppose that they are remarkably unusual. From population modelling and studies of the wasp behaviour and biology, Hochberg (2000) suggested 'Conservation guidelines for the parasitoid need be only slightly more stringent than those for the butterfly'. He recommended particular attention in this case to assuring the wellbeing of sites harbouring productive colonies of the specific host ant, *Myrmica schencki*.

More generally, Shaw (1990) noted that some conspicuous groups of hymenopterous parasitoids 'appear to have suffered massive declines in western Europe' in the second half of the twentieth century, but also that most groups of these wasps are too poorly known for any such assessment to be made. The importance of preserving for taxonomic study all parasitoids reared from insects in captivity, particularly from hosts of conservation interest, is considerable. Only by such efforts can these parasitoids be documented effectively, because retrieval of host records from published accounts is often problematical. Shaw (1990) noted the following difficulties of relying on such information:

1. Data are usually not quantitative, so that equal weight is given to regular and 'freak' associations.
2. Many published names currently have uncertain status, with much synonymy unresolved.
3. Parasitoid identification errors are extremely common, reflecting poor taxonomic knowledge and the existence of many 'species aggregates' which may have been understood differently at the time records were published.
4. Host misidentifications are 'surprisingly numerous', and often radically wrong because the true host was overlooked (such as by being introduced with food plants).
5. Usually, no distinction is made between primary and secondary parasitoids (hyperparasitoids).

Specialist-accumulated and curated reference collections are a major source through which such problems may be progressively overcome. At present, the levels of 'misinformation' that may result from non-specialist interpretations of published information are likely to be both high and of minimal use in promoting effective conservation measures.

Community modules and insect species conservation

'Community modules' (*sensu* Hochberg *et al.* 1996) comprise the small number of intimately interacting species whose interactions can be understood and are reasonably isolated from much of the rest of the community in which they occur. They may need to be conserved together as forming obligatory or near-obligatory mutualisms or interdependencies. They may, for example, consist of an insect, its sole food plant, any host-specific parasitoids or monophagous predators, and, possibly (as in some butterflies), a mutualistic ant. In turn, the plant may have a specific pollinator, whose loss would be catastrophic. Any or all of these may be threatened and that state influence the other players involved. The practical need is then to conserve these interacting species together as modules that broadly parallel small food webs and emphasise the frequent need for extending focus beyond the immediate realm of the target insect. This concept stems from conservation of *Maculinea* butterflies, their mutualistic ants, and their specialised ichneumonoid parasitoids (p. 46) and the context was discussed in detail by Mouquet *et al.* (2005).

This system of interacting species had been studied for more than 20 years, and the particular features influencing the conservation of *M. alcon* and its host plant *Gentiana pneumonanthe* and their long-term persistence were reasonably clear. Models (Box 1.5) to simulate the effects of four conservation management strategies (burning, sod cutting, mowing, grazing) showed that the mechanisms optimising size of the *Maculinea* and *Gentiana* populations differed, so that choices must be made in pursuing the best conservation balance. Mouquet *et al.* (2005) argued that the precise conservation measures needed will be determined by different field conditions, so that management must be pursued on a case-by-case basis to reflect different site and population features.

Box 1.5 · *Modelling a community module for an insect: traditional land uses in butterfly conservation*

The monophagous Alcon blue (*Maculinea alcon*, see also Box 1.3) associates obligately with the larval food plant (*Gentiana pneumonanthe*) and the ant *Myrmica scabrinodis* (over much of its range), with these three species forming a community module whose integration depends on all three being present. Management of the butterfly necessitates conservation of both the other species involved. The module's persistence has depended on human activities over some 5000 years, so that traditional forms of land use for agriculture (burning, grazing, mowing) or for fuel extraction (cutting peat sods) have been applied to its habitats over a long period. Effects of these practices were considered in a simulation study by Mouquet *et al.* (2005, see text) to explore possible conservation outcomes. The approach and assumptions are displayed in Table 1.5.1, with several different regimes for grazing (strong, intermediate, weak), mowing (5 cm, 10 cm) and sod cutting (5%, 15%, 25%), with the major results shown in Table 1.5.2. With additional simulations of mixed treatments, predictions were made of persistence and population sizes of these species, to compare different management regimes.

Regular burning produced the highest densities of all three species, with periodic grazing providing the next highest densities of butterflies and ants and adequate numbers of gentians. Mouquet *et al.* recommended that regular winter burning, largely abandoned, should be gradually reinstated on heathlands. Sod-cutting has been employed

Table 1.5.1 *The conservation strategies and conditions used in the above simulation models*

Abbreviations: ts, time since the last perturbation; Sd, percentage of area used in sod-cutting.

Strategy and taxon	Effect
Burning (ts is set to 0 the year after burning)	
Gentiana	Seedling survival is set to 0 the year of the fire; juvenile survival is reduced by 50% the year of the fire.
Myrmica	In peat bogs, 755 of nests destroyed.
Maculinea	In peat bogs, caterpillar survival within ant nests reduced by 75%.
Grazing (with periodic grazing, ts is reduced by 75% if intensive, 50% if intermediate and 25% if weak. With year-round grazing, ts is relatively stable: 2 if strong, 4 if intermediate, 6 if weak)	
Gentiana	Adult survival reduced by 10%, seedling and juvenile survival, and adult fecundity reduced by 60% (intensive), 30% (intermediate), 10% (weak).
Maculinea	Survival on plants reduced by 60% (intensive), 30% (intermediate), 10% (weak).
Mowing (ts reduced depending on height: 2 at 5 cm, 4 at 10 cm above ground)	
Gentiana	5 cm: juvenile survival reduced by 25%, adult survival reduced by 5%. Seedling and juvenile survival reduced by 10% by trampling.
Myrmica	5 cm: ant growth rate reduced by 25%.
Maculinea	5 cm: ant searching area and number of caterpillars adopted per nest reduced by 25%.
Sod-cutting (ts reduced by 30% if Sd 55, 60% if Sd 15%, 80% if Sd 25%: reduction in ts occurs one year after a sod-cut)	
Gentiana	Seedling, juvenile, reproductive adult and dormant survival reduced by Sd.
Myrmica	Number of ant nests reduced by percentage area removed, Sd.
Maculinea	Caterpillar survival in ant nests reduced by percentage area removed, Sd.

Source: after Mouquet *et al.* (2005).

successfully, and is an easily controlled method for use on sites on which grazing is impracticable. Last, regular mowing can produce large numbers of flowering *Gentiana*, but results also in low numbers (or absence) of ants and butterflies. Mixing different strategies may lead to persistence of the module for less overall effort.

Table 1.5.2 *Effect of different management strategies on the mean population sizes of the three species*

Simulations run for 100 time steps; figures are butterflies per hectare, ant nests per hectare, plants per square metre; dashes indicate that there was no effect of treatment.

			Years			
Strategy and species	2	3	4	6	10	15
Burning						
Gentiana	0	2.5	3.6	3.65	2.84	0.016
Myrmica	1100	302	242	160	30	0
Maculinea	0	952	755	448	0	0
Grazing (strong, intermediate, weak in vertical sequence)						
Gentiana	0	0.219	1.96	1.47	0.032	0
	1.71	1.74	0.86	0.022	0	0
	0.3	0.005	0	0	0	0
Myrmica	982	619	184	33	0	0
	214	67	0	0	0	0
	0	0	0	0	0	0
Maculinea	0	664	480	0	0	0
	602	111	0	0	0	0
	0	0	0	0	0	0
Mowing (5 cm, 10 cm, in vertical sequence)						
Gentiana	3.46	3.58	3.48	2.9	1	0.015
	–	–	1.97	1.01	0.048	0.0008
Myrmica	114	92	66	3	0	0
	–	–	0	0	0	0
Maculinea	252	198	120	0	0	0
	–	–	0	0	0	0
Sod-cutting (5%, 15%, 25%, in vertical sequence)						
Gentiana	1.15	0.049	0.0018	0	0	0
	1.81	1.59	0.908	0.05	0.002	0
	0	0.72	0.84	0.37	0.0037	0
Myrmica	0	0	0	0	0	0
	102	40	0	0	0	0
	616	108	64	0	0	0
Maculinea	0	0	0	0	0	0
	260	43	0	0	0	0
	0	274	125	0	0	0

Source: after Mouquet *et al.* (2005).

The occurrence of such modules, and their contribution to understanding both the scope and the detail of conservation, is an important consideration. By their very nature, community modules may be a valuable tool in extending conservation interest and capability, for example by superimposing botanical interest in threatened plants on the primarily more isolated entomological field of endeavour and, more widely, emphasising the roles of such critical resources for the focal insect species.

Summary

1. The enormous number of insect species renders focusing on the most deserving individual species difficult. Resources and support are grossly inadequate to deal with all deserving cases, and priorities have to be set.
2. Risk of extinction is one criterion used frequently to accord such priority (as a form of triage), and links with IUCN criteria for categorisation of threat, with these echoed in much protective legislation for insects in many parts of the world.
3. Lists of threatened insect species for any region are almost invariably not comprehensive, or even reasonably representative of the true extent of conservation needs. Additional criteria for selection of species for conservation include taxonomic position, ecological factors, vulnerability (such as by documented decline in numbers and/or distribution), public support, and others, all of which can be applied to either select or rank species for priority treatment. Although cited often, the term 'rarity' does not necessarily demonstrate or correlate with conservation need. Definition of threats to an insect provides a basis for the management of the species.
4. Legal designation as 'protected' or some similar status can oblige further investigation at local or wider scales.
5. Categories of threat for insects can not adequately incorporate population data or quantitative estimates of probability of extinction: simply, such information is almost invariably not available. Many insects are very difficult to count in the wild, because their activity may be highly seasonal, and subject to vagaries of weather and factors such as variable diapause that can further affect apparency. Many species also undergo normal numerical fluctuations of severalfold in abundance in successive generations, rendering many inferences of genuine decline tentative, particularly without long-term observations.

6. Many insect subspecies or 'significant populations' (collectively sometimes referred to as 'evolutionarily significant units') remain the subjects of taxonomic uncertainty over their precise status. Allied to this,
   inadequate taxonomic knowledge for many vast groups of insects renders recognition of species impossible other than by a few specialists,
   so these insects are not tractable for species-level conservation treatment. Most insect species conservation has focused on better-known
   or more popular groups. Amongst these, butterflies are particularly
   important for their generally favourable public image.
7. Extreme ecological specialisation and relationships between species in
   assemblages has led to promotion of considering 'community modules', so that species that form obligate relationships may be considered
   together.

2 · *Plans for insect species conservation*

Introduction: basic principles and scope

Conservation management for a threatened species, be it insect or other, has two universal aims:

1. In the short term, to minimise or eliminate its risk of extinction, by removal of threats and increasing its security.
2. In the long term, to provide conditions under which the species can continue to thrive and to retain its potential for evolutionary development, ideally without continuing intensive (expensive) management.

Most attention is given to the first of these objectives, and this is the only one for which most current management plans cater effectively. Species-focused conservation plans, under names such as recovery plans, action plans, action statements, management plans, or some other similar epithet, have been produced as components and drivers of numerous insect conservation programmes. These varied titles imply rather different themes, but contents of the documents overlap considerably in practice, and titles of some may simply reflect the specific wording in different governing regulation or legislations, and the depth of the treatment in the documents that flow from these. And, indeed, the scope of the document may be dictated in principle by the governing legislation under which an insect is listed, with very specific requirements sometimes given. Whatever the name, these documents signal that the focal species has/have in some way been selected or singled out for conservation need or consideration at some level, to promote either protection from decline and loss, or recovery from earlier such losses and to reduce vulnerability for the future. Most commonly, such plans flow from formal listing of the species as 'threatened' or 'protected' in some way. Others stem from documents prepared earlier for nomination of insects for such considerations. Yet others may arise from the zeal of individual proponents. Plans vary widely in scope, length and complexity, but 'a plan' is a prerequisite for any focused conservation management exercise, to specify and guide that management.

The sequence of basic steps in an insect species management plan (New 1995) is shown in Fig. 2.1. The initial need to critically establish conservation status (as a dynamic categorisation subject to change to either greater vulnerability or greater security) determines whether a species is threatened, secure, or may need to be formally 'listed' or otherwise merit individual conservation attention. The threats inferred or detected may be associated clearly with local losses or declines, or no such direct evidence of their effects may be detected – often for lack of previous monitoring or other earlier information to provide a sound baseline against which to evaluate changes. The causes of any declines found, in either or both of range and abundance, dictate the kinds of management needed to mitigate the threats, and their extent dictates whether recovery is needed or whether simply removing the threats and safeguarding what is left effectively may be the better option. Monitoring (p. 196) is needed as a basis for evaluating any management undertaken, and to guide its possible modifications. Threats may affect individuals directly (for example, by killing them or affecting their behaviour or reproductive capability) or the habitat or resources on which they depend. Management has both regulatory and scientific components, the latter usually paramount but often facilitated by the former. Both are integral components of many insect management plans.

Using this rationale, New *et al.* (1995) summarised planning for insect species conservation into a four stage sequence: (1) status evaluation; (2) delineation of threats and their likely effects on habitats and individuals; (c) definition of the remedial strategies; and (d) implementation of management to control threats, protect and/or restore habitats and conserve the species. In a similar mode, McGuinness (2007) outlined what he thought of as 'a simple process' for insect species conservation plans, to help focus among the numerous species needing help. The key questions he suggested should be posed are:

1. Which species do you want to protect? – that is, to set priorities among the numerous deserving candidates, not all of which can be supported individually.
2. What is causing them to decline? – that is, to determine the prime motivation for their conservation.
3. Can we manage the agents of decline?

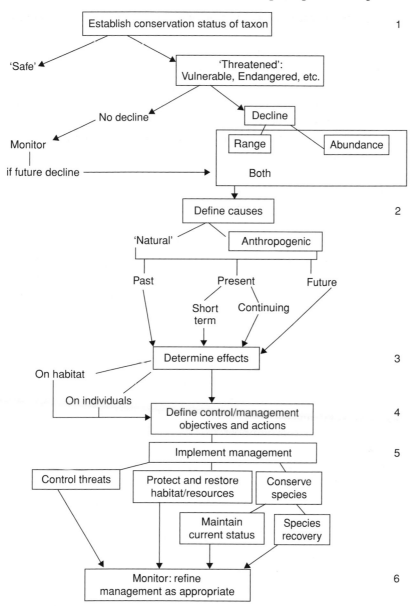

Fig. 2.1. A sequence of steps in a single species conservation plan for an insect (after New 1995).

4. What are the key sites? – again, a prioritisation exercise, with selection incorporating other values such as finding the areas that will conserve the greatest numbers of threatened species.
5. Does it constitute a priority for funding?

This pragmatic approach is summarised in Fig. 2.2. It raises the difficult, sometimes emotional, questions of how to set priorities at the three practical levels of species, sites, and funding and related support for conservation.

Of these, 'sites' may be the easiest to address. Much insect conservation focuses, at least initially, on particular sites on which the species is found and where it is believed to be threatened. For many insects, choice between sites to be conserved may not even exist (with the species known only from single sites), or be otherwise extremely limited in distribution. Should choice be needed, parameters may include size of the site (more broadly 'habitat patch') (larger better than smaller), degree of isolation (is it/can it be connected functionally to other occupied sites, perhaps as a metapopulation module, or is it more isolated?), extent of degradation or naturalness ('more natural' sites better, reflecting the costs and intensity of likely management, such as restoration of degraded sites), security (increased nominally if it is in a reserve, or can be protected easily, preferably without the large costs involved in land purchase), ease of monitoring (close to a home base), and additional conservation values (other threatened species present, or other definable values) so that its conservation may garner support from elsewhere. Selected sites must be safeguarded as effectively as possible, and their extent and position defined by tools such as GPS mapping (augmented by overlays of critical resources: p. 83) and aerial photographs.

As noted in Chapter 1, grounds for setting priorities among species are very varied, and can become subjective and responsive to the zeal of individual proponents. Both absolute and relative priorities may apply, and the rationale for setting these will continue to be debated within the general consensus that the 'most deserving species' from among those under consideration should receive priority. A common necessity in order to obtain funding, though, is that the species be recognised generally as deserving conservation, for example by formal 'listing' on some national or regional schedule of 'threatened species' or 'priority species'. In many places, this or equivalent acknowledgment is a 'passport condition', because only then may the species be granted official recognition of need and be eligible for funding. Nevertheless, numerous species may

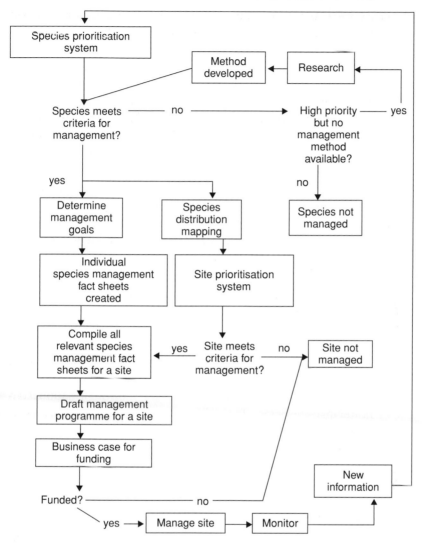

Fig. 2.2. A decision-making process for a scheme for site management, based on insect species conservation (after McGuinness 2007, with kind permission of Springer Science and Business Media).

gain this status, as an initial hurdle toward acknowledging conservation need. Most insect management plans are for species selected in this way.

Occasional additional hurdles may arise in the 'listing process'. For example, not all nominations for consideration may be allowed to proceed expediently. Thankfully, this is rare, but it may mean that a given species may never be accorded the level of formal consideration needed to ensure that its conservation needs may become 'official'. For Australia's EPBC Act, nominators are advised that 'although all nominations are welcome, it may not be possible to include all nominations on the Proposed Priority Assessment List (PPAL) in any one year'. In formulating the PPAL, the Committee may consider a range of factors, including (1) the level of threat to the species, (2) the role the species plays in ecosystem function, (3) the benefits of listing the species, for example in terms of legislative protection and threat abatement, (4) the capacity to effect recovery of the species, and other factors. The limitation on numbers of nominations allowed to proceed is considered in relation to the workload of the committee, and it is of limited consolation to read 'To balance its workload, nominations not included in the PPAL may be considered by the Committee for prioritization in the subsequent round' (DEWR 2007).

Scales and focus

Recovery plans 'are the central documents available to decision-makers and serve as guides for the management and recovery of threatened and endangered species' (Boersma *et al.* 2001). Collectively, they have been prepared for substantial numbers of different taxa, in many parts of the world, and to apply at scales ranging from global, through national, to local or municipal scenarios. They differ enormously in length, scope and complexity and range from rather superficial or bland 'motherhood statements' of mission or good intent to detailed practical analyses of conservation needs and how to fulfil them. They can also include summaries of all relevant biological information. Plans may deal with one or more species, either in isolation or in the wider contexts of habitat, community or site conservation needs.

Recovery plans may be prepared by agencies or contractors, with varying levels of peer review or wider comment sought before their adoption. The centrally important need is to incorporate the best available expertise on the taxon/a treated and apply this knowledge to devise and undertake the best possible conservation management. However, plans made by an informed 'recovery team' (p. 214) are likely to incorporate the most complete and up-to-date advice, and this approach of expert input has been advocated widely (Burbidge 1996). Thus, the recovery plan for the Karner blue butterfly (*Lycaeides melissa samuelis*) was prepared by the butterfly's Recovery Team for the United States Fish and Wildlife Service (KBBRT 2001).

Here, some aspects of these documents developed for insects in various parts of the world are examined briefly to demonstrate how they operate in practice, as well as how they might operate better.

Most such plans target single species. Those that include a group of related taxa are necessarily more general in broad scope, but usually incorporate individual considerations for each species, following wider initial comment and perspective. At this level, they intergrade with broader documents such as Red Data Books (which identify species needing conservation and promote their conservation) but may differ in practice in committing to action rather than being simply advisory in nature. Multispecies recovery plans may be invaluable in helping to define general protocols on which to found conservation action for any included species, as well as being attractive to agencies that produce them by providing information applicable across an array of taxa. The 'Threatened Weta Recovery Plan' for New Zealand (Sherley 1998) is one such case, based on overlapping aspects of the biology of included species. Other multiple plans may include treatments for 'better known' and 'poorly known' species, as in that for New Zealand Carabidae (McGuinness 2002).

Wider regional Action Plans are exemplified by that for *Maculinea* butterflies in Europe (Munguira & Martin 1997). That group of five species of large blues (Lycaenidae) has collectively attracted, perhaps, more recent conservation attention than any others of this popular family of butterflies. All these species are of considerable conservation concern, and are important 'flagships' for the discipline of butterfly conservation in Europe. A major EU-funded project has recently led to substantially increased perspective of the butterflies and their management needs, with some early results summarised by Settele *et al.* (2005). Thus, although *Maculinea arion* occurs in more than 30 countries, its extinction and subsequent re-introduction into the United Kingdom (p. 184) is one of the most intensively appraised cases of insect conservation. The five species differ in details of their biology, but also have strong unifying features in having mutualistic relationships with ants of the genus *Myrmica* and in having declined through loss of habitats from agricultural intensification and abandonment of previously grazed or mowed meadows. These changes led to alteration of the grassland dynamics and the suitability of areas for both larval food plants and for the individual species of *Myrmica* ant with which each *Maculinea* associates in any given part of its range. Habitat changes as above were regarded as 'global threats', upon which more specific local threats were usually superimposed. The European Action Plan used data from 13 countries to assemble a conservation strategy for *Maculinea*. The status of each species in each country was evaluated, together with appraisal of threats, and a list of conservation actions for each species in each country from which information was available was provided.

In contrast to the wide scale of this international approach, much planning for insect species conservation focuses on very local issues resulting from changes to individual terrestrial or freshwater sites, and the concerns of local naturalist groups for notable species recorded from these. Many cases are founded in 'crisis management', resulting from plans to develop particular sites on which notable species occur, or to change the site in some way, such as by road construction or impoundment. Planning then becomes essentially a local or municipal issue, but may subsequently take on broader implications or relevance. This municipal level of conservation has massive importance in initiating and fostering interest from the local community. As Australian examples, initial local concerns for the Eltham copper butterfly near Melbourne (Braby *et al.* 1999; Canzano *et al.* 2007) and the small ant-blue butterfly at Mount Piper (New & Britton 1997), both in Victoria, led to adoption of these species as significant local emblems. Their conservation needs were embraced by local communities and the local authority, without whom the respective conservation plans could not have been promoted effectively.

Many such local exercises, involving the 'grassroots' level of conservation interest, may predominantly involve fostering the interest and goodwill of non-scientists, such as of other people concerned for the welfare of the local environment and who are seeking general issues in which they can become involved toward that end. Insect conservation still has 'novelty appeal' in many places. Local newsletters, meetings and the formation of local 'friends' groups' may then to some extent replace formal management or action plans, but the latter may still be invaluable in providing biological understanding and impetus to drive conservation effectively, and effective coordination of activities is important. Thus, a 'Local Species Action Plan' for the peacock butterfly (*Inachis io*) and other butterflies for the Royal Borough of Kensington and Chelsea (London) (RBKC 1995) has three aims toward fostering wider awareness of the local environment, using the peacock as a focus, as follows.

1. To develop suitable habitats and to encourage and increase the populations of the peacock butterfly and other butterflies in Kensington and Chelsea.

2. To increase public awareness and understanding about the peacock butterfly and other butterflies, and to encourage the local community to participate in 'butterfly gardening' to benefit butterflies.
3. To collect butterfly data to monitor their distribution and as an indicator of the quality of the local environment.

Both direct (peacock) and wider (other butterflies, local environment) measures are encompassed in this series, with some additional more specific measures endorsing these in greater detail. The simple but pertinent contexts of such local plans are an excellent public relations exercise, and merit consideration as components of many insect conservation plans of wider geographical relevance and for which local practical focus may contribute to, and augment, wider operations. Public interest and involvement is often itself a critical resource in insect conservation (p. 214). For the Richmond birdwing butterfly in Australia, Sands *et al.* (1997) involved numerous school groups in planting food plant vines for caterpillars, and monitoring their condition: a novel 'Adopt a Caterpillar' scheme involving a number of schools also attracted wide attention. The conservation programme (p. 227) for this spectacular butterfly is supported by the widespread Richmond Birdwing Recovery Network, launched in 2005 and which, among other activities, promotes liaison between community members and relevant government authorities. The conservation of *O. richmondia* was earlier (from 1999) based on a community conservation project (Sands & Scott 2002), with the following aims.

1. Identifying and protecting natural habitats for *O. richmondia*, including investigating conservation management agreements for private properties, land acquisition by local authorities, and involving the community in surveys.
2. Mapping and recording natural sites for breeding of *O. richmondia* and its larval food plant.
3. Replanting and enriching vine communities, to link existing colonies and extend distribution to cover its former range.
4. Creating signposting for vine sites to raise community awareness and aid site protection.
5. Identifying plant communities associated with the birdwing's larval food plant vine.

Numerous authors have advanced their ideas for what insect management plans should comprise. Inevitably there is considerable variety and difference in emphasis, reflecting the great variety of places and

Table 2.1 *Arnold's (1983a,b) proforma for species-orientated conservation of lycaenid butterflies*

1. Preserve, protect and manage known existing habitats to provide conditions needed by the species.
   (a) Preserve; prevent further degradation, development, or modification
       i. cooperative agreements with landowners and/or managers
       ii. memoranda or undertakings
       iii. conservation easements
       iv. site acquisition (purchase/donation of private land) or reservation (public land)
   (b) Maintain land and adult resources
       Minimise threats and external influences
   (c) Propose critical habitat
   (d) If recovery, clarify taxonomic status of taxon in habitat and other populations
2. Manage and enhance population(s) by habitat maintenance and quality improvement, and reducing effects of limiting factors.
   (a) Investigate and initiate habitat improvement methods as appropriate
   (b) Determine physical and climatic regimes/factors needed by species and relate to overall habitat enhancement on site
   (c) Investigate ecology of species
       i. lifestyle and phenology; dependence on particular plant species or stages
       ii. dependence on other animals, and their roles
       iii. population status
       iv. adult behaviour
       v. determine natural enemies and other factors causing mortality or limiting population growth
   (d) Investigate ecology of tending ant species, if present
   (e) Investigate ecology of food plant species
3. Evaluate all the above and incorporate into development of long-term management plan. Computer modelling may assist in making management decisions.
4. Monitor population(s) to determine status and evaluate success of management
   (a) Determine site(s) to be surveyed, if choice available
   (b) Develop methods to estimate population numbers, distribution, and trends in abundance
5. Throughout all of the above, increase public awareness of the species by education/information programmes (such as information signs, interpretative tours, audio and visual programmes, media interviews, etc).
6. Enforce available regulations and laws to protect species. Determine whether any additional legal steps are needed, and promote these as necessary.

circumstances in which insect species conservation is needed or contemplated. Each clearly has its merits within its primary context, and it is perhaps futile to even seek wider protocols and generality. However, in practice, some plans 'work' better than others, and numerous elements are recurrent in many different plans. Arnold's (1983a,b) proposed management scheme for lycaenid butterflies in North America (Table 2.1) contains many of the practical themes that may need to be considered in any equivalent plan for an insect and, indeed, that *have* been reiterated in numerous later exercises of this sort. It includes the various strands of communication and agreement needed to assure site security and management, needs for biological information on a suite of issues, and the importance of monitoring and education, with any or all of these facilitated by legal means where possible and necessary.

New Zealand's Department of Conservation differentiates formally between 'Recovery plans' and 'Action plans'. Action plans have no formal definition, but are written with a general standardised format and provide broad detail of species distribution, threats, future management needs, and requirements for research, survey or monitoring (McGuinness 2007). They serve a valuable advocacy role, in helping to raise awareness of the plight of taxa. Recovery plans, so far available for only three groups of New Zealand insects (weta, carabid beetles, short-horned grasshoppers), are statements of intent that focus on goals and objectives over a defined period (usually of 5–10 years), and require annual reporting of progress with duties delegated to suitable staff for implementation. This approach to New Zealand Carabidae (McGuinness 2007) demonstrates that knowledge of the 56 species considered is highly uneven. Four species were selected for individual recovery plans, but most beetles were simply categorised into one or other of several broad groups as (1) beetles requiring survey or other information-gathering, (2) beetles requiring taxonomic clarification, and (3) beetles listed previously as of concern but now downgraded and with no conservation actions proposed. The major research needs for these groups were thus seen as development of effective survey methods and taxonomic revision of several complex genera.

Management options

The ambit scope of an insect recovery plan can, in principle, be very wide and range from little action being needed anywhere within the species' range to intensive and interventionist management for a species across its entire range. This variety is demonstrated clearly for weta

(Sherley 1998), for which the following four management options were specified.

1. Do nothing. Many populations were considered to be able to survive, and some island populations to remain abundant without management, but others would be likely to become extinct. More widely, to 'do nothing' is an important conservation management option for insects, where the risks of 'doing something' in a climate of ignorance need to be considered very carefully. Doing nothing is then a positive management decision.

2. Manage selected populations and their habitats. This involves a full suite of management options for selected populations which are given priority over others. Such 'priority populations' in Sherley's account include those with a high level of genetic or morphological variation; the sole remaining populations of species; populations decreed important from an ecological perspective; or populations important numerically. This approach may lead to extinction of some populations but security for the species as a whole.

3. Manage all populations and their habitats. This involves measures as for option 2, but may involve higher costs, depending on the total number of populations and sites treated. The outcome of this approach could be success but, if costs not be sustained, could entail risk of a higher chance of failure because efforts become 'diluted' through lack of support.

4. Establishment of multiple populations of each species. This approach involves 'spreading the risk' to each species through translocations to establish additional populations on suitable sites, perhaps necessitating continuing site management. Sources of insects are twofold; from secure wild populations or from captive-reared stock. Should establishment occur, the additional populations should help to ensure long-term survival of the species involved.

Essentially, insect recovery plans have three major purposes, with priorities among these differing with context and constituency. These intergrade in many ways but, broadly, are (1) as 'appeasement' to fulfil, simply by their production, legal obligations conferred by listing the species in some formal way; (2) as public relations exercises, with importance in increasing awareness of the parlous plight of species and fostering commitment to their conservation; and (3) as comprehensive critical summaries of conservation need, and of the steps needed for effective practical conservation progress to be made. Each of these may be viewed, at some level, as a facilitator for conservation to progress, but the needs visualised for a 'more political' document may differ considerably from those of one intended to dictate and drive practical conservation management. At the extreme, a Minister (or other authority) may seek to fulfil an obligation by simply producing 'a document', with little real intention to translate it to reality, simply to be seen to be 'doing something'. In such, fortunately increasingly limited, contexts, quantity (number of documents or species dealt with) may be more important than quality or scientific integrity and practicability. In many legislations, a Minister is obliged to seek advice from, or consult, a scientific advisory committee, about ways forward for threatened species, but not necessarily to heed that advice. The extent of such formal obligations varies widely across different legislations, but there is widespread community expectation that practical conservation 'actions' should flow from any published management plan.

Taking any such action depends on availability of support, predominantly of expertise and funding, both of which are commonly in very short supply. The capacity to produce sound plans for insect recovery is very limited in most State or Territory agencies within Australia, for example, and a major recommendation by Yen and Butcher (1997, and echoed by Sands & New 2002) that an 'invertebrate expert' be appointed to each such body has not yet been entirely fulfilled. Parallel gaps in expertise at those levels are numerous. Outcomes of this lack are that plans for insects (and other invertebrates) must often be drafted by people versed in vertebrate biology alone and against a background of threat criteria interpreted as for relatively well-known vertebrates. A practical consequence is that such plans may be given only low priority in relation to others that such people feel more confident in producing and taking forward into practice. McGuinness (2007) referred to this as taking people 'out of their comfort zone or skill set', and noted that it may result in reluctance to act 'for fear of doing something wrong'. This is not a trivial concern, but flows also from the generally poor public image of many insects and the belief that they may be conserved adequately under the umbrella of other species, such as larger vertebrates, for which public sympathy (and professional knowledge and confidence to act) is much higher. Likewise, the pool of consultants available to draft such plans for insects, or to review drafts, may be very limited, particularly if those people involved in promoting the particular insect species earlier for listing are excluded as interested parties. Often, they are the only people with firsthand field knowledge of the species and its putative plight. Nevertheless, wide consultation is common in drafting insect recovery plans, but most such input and peer review involves the quality of information presented, rather than the feasibility of implementation. Recovery teams, convened or appointed to oversee plans, normally include representatives of all interested parties on whom responsibility for the plan will devolve, and one or more independent advisors.

Lack of 'insect expertise' leads to serious considerations over undertaking management in terms of 'who does what', especially when there is very limited in-house logistic capability. For swallowtail butterflies, New and Collins (1991) suggested four possible avenues to help counter this. The first three, in particular, have much wider application in insect species conservation.

1. Direct employment of (independent) scientists to do the work.
2. Liaison with local scientific/naturalist groups, with provision for funding to coordinate searches (etc.) and prepare reports.
3. Funding of postgraduate studies through local universities.
4. Possibility of 'ranching' as a major practical 'spin-off' from other conservation activities for swallowtails. There are concerned entrepreneurial lepidopterists in many parts of the world, and support for ranching (added here: and developing captive breeding programmes) seems to be warranted.

Sound biological knowledge and understanding is a key element of any such plan, and integral to formulating both objectives and actions. In North America, Schultz and Hammond (2003) noted that the United States Endangered Species Act (ESA) demands 'objective, measurable criteria' on which to base listing decisions, and this is fundamental also in management plans. Few conservation biologists would query the need for the best possible scientific information to underpin any recovery or other management strategy.

However, Schultz and Hammond (2003) reviewed 27 recovery plans for insects listed under ESA, and showed that stated recovery criteria were usually linked very poorly to species biology, by lacking quantitative recovery criteria linked to biology. Recovery criteria stated in those plans were placed into four categories.

1. Specified minimum population size or growth rate.
2. Specified duration for minimum population size or growth rate.
3. Metapopulation criteria.
4. Criteria regarding permanent habitat protection and/or management.

For minimum population size and growth rate, four types of criterion were delimited: no criteria, a 'self-sustaining population', a specified minimum population size, or a stable or increasing population growth rate required. The last two of these were considered to be quantitative.

Many other authors, also, have endorsed the essential and central role of scientifically based recovery criteria (Clark *et al.* 2002; Gerber & Hatch 2002, both on ESA). In the absence of defined measurable criteria, 'recovery' is commonly projected on the wellbeing of sites on which the species resides, as the best interim measure for conservation. Thus, in treating five poorly known species of *Synemon* (sun-moths, Castniidae) in Victoria, Douglas (2003) specified a number of intended management actions to increase site security and prevent further degradation. Intersite variations across a species' range may, indeed, demand this approach (p. 220).

Lack of detailed knowledge, however, is frequent and the problem then arises as to how to treat the numerous poorly known insect species in individual conservation management. A significant policy pointer from the New Zealand carabid recovery plan (McGuinness 2002) arises from the fact that all the four individual species selected for recovery plans are incompletely known, and concerns for their decline remain to be scientifically justified. However, action is being taken now, on the precautionary basis that delay may lead to demise. The ability and willingness to act in this way, based on 'educated guesses' of need, and initial appraisals of likely threats and causes of decline, deserves widespread consideration for adoption elsewhere. For many insects, knowledge of their real conservation status and needs has come only after formal notice has been made, and 'listing' has facilitated or stimulated the investigations needed to clarify this. In some cases, the exercise may flow from discovery of a species for which conservation is urgent (for example because of imminent development of the site on which it was found), and a rapid plan (perhaps accompanied by an interim conservation order or moratorium on development of the site) is needed as a temporary measure, pending development of a more informed document at a later stage. Recommendations of such documents are inevitably rather general, but it is important that the compass of possible conservation needs be anticipated as far as possible. For the Eltham copper butterfly, the first action plan (Vaughan 1988) produced soon after the butterfly was rediscovered near Melbourne projected three main aims.

1. Protection of the colonies from threatening processes associated with urbanisation.
2. Provision for increase in effective habitat by promoting natural regeneration of *Bursaria spinosa* (the sole larval food plant) and propagation

from seeds or cuttings, or by transplanting from any sites to be sacrificed for development.
3. Provision for a ranger to foster and undertake practical management.

These thereby addressed (1) threat evaluation and abatement, (2) habitat security and enhancement, and (3) need to coordinate and monitor management. With such elements assured, more informed management can be developed from a relatively secure basis or, at least, a considered 'holding operation' likely to provide at least some practical benefit. Any such 'interim conservation plan' is invaluable, but there is some danger that it may become 'permanent' unless impetus is sustained.

At times, it may be necessary to formulate a broad initial 'mission' for a management programme to convey a suite of values, including ideals and topics such as influences on human welfare, to a wide audience. For the large international conservation programme for Queen Alexandra's birdwing butterfly (*Ornithoptera alexandrae*) in Papua New Guinea, the 'conservation mission' includes the following overall rationale: 'To ensure the survival of the remaining *O. alexandrae*, through a commitment to conservation which involves other improvements to the welfare of conservationist/landowners and their neighbours; which raises the possibility of ecotourism; and which at least postpones exploitation until resource extraction, resource management, returns to landowners and decision making by landowners are improved' (Anon. 1996, p. 14). Five major components were included in this plan.

1. Research, to enhance understanding of the distribution, biology and ecology of *O. alexandrae*.
2. Conservation of Primary Habitat Areas to maintain the existence of all important primary habitat areas.
3. Education and awareness: to promote knowledge of and concern for *O. alexandrae* throughout the country.
4. Economic and social issues: to provide economic and social incentives and measures for conserving *O. alexandrae* habitat.
5. Project management; to coordinate and manage inputs and implement activities (AusAID 1999).

These two butterflies represent extremes in conservation perspective and potential. On the one hand, the Eltham copper is a small lycaenid occupying very small isolated sites in an environment sympathetic to butterfly conservation and with resources to undertake such exercises. It typifies many of the species treated in this book. On the other hand, the world's largest butterfly occupies remote areas of northern Papua New Guinea and ranges across wide, largely inaccessible landscapes of tropical primary forests in serious need of protection from loss to logging and oil palm plantation. Pressure to conserve *O. alexandrae* (notwithstanding its being a 'national butterfly' of PNG) emanate largely from outside the country, and the resources needed are mostly not available without international support. The case is discussed further on p. 209 but human values and interests are included firmly and prominently within the ambit of a conservation management plan. Without these considerations, efforts to conserve *O. alexandrae* would be likely to be largely futile.

Conservation on single small sites and across whole landscapes clearly demand rather different perspectives, but many of the management objectives (and actions flowing from these) are in common at these differing scales.

A third class of scale occurs with island endemic species. The arena of concern may be the whole island, as the maximum natural range of the insect, but more detailed focus usually occurs on specific sites. Thus the spectacular Jamaican endemic swallowtail butterfly *Papilio homerus* was formerly widespread but now occurs only in two small populations on the island (Emmel & Garraway 1990; p. 210); the giant Frégate island tenebrionid beetle (*Polposipus herculeanus*) is another example (p. 211).

Assessing progress

The task of practising insect conservation biologists is to bring the conservation undertakings made in such plans to fruition, sometimes in complex political arenas, and it behoves us to 'interfere' and influence these plans as constructively as possible to ensure that their objectives are sound, sensible and feasible. Objectives must be enunciated very clearly, not least to assure optimal effect and progress, and as a prelude to determining actions.

Following the broad objective (or 'mission') of the document, a listing of compartmentalised objectives commonly occurs. The need for 'SMART' objectives reflects that each objective should be Specific (unambiguous), Measurable (with criteria and timing for this stated), Appropriate (related to the long term over-arching goal of the plan), Realistic (achievable within the time frame specified), and Time-bound (with a cut-off date for attainment). Some workers replace 'Realistic' for 'R' in the above acronym with 'Responsibility', to designate which agency or person will commit to the task. The objectives stated in many published insect recovery plans fall far short of meeting all these criteria. The last (Time-bound) is particularly important in assuring commitment to action (so that the plans are not simply shelved) and, perhaps, is that most frequently not addressed. Many insect recovery plans include a stated 'review by' date, which (in common with those for other groups) may not be met because of logistic limitations or changed priorities. Linked with this, monitoring of progress is critical both to determine success and to render management adaptive and responsive to changing circumstances; a recovery plan should not be inflexible. Responsible review ensures that additional information will indeed be incorporated, and that the plan is dynamic. The converse is that an unreviewed plan will in time become suboptimal or, even, misleading.

The initial objective of a species recovery plan may thus at times have the flavour of a 'mission statement', rather than be simply a biological one. The plan for the Schaus swallowtail butterfly (Ssb, *Heraclides aristodemus ponceanus*, p. 24) states 'The objective of the 1999 Schaus swallowtail recovery plan is to reduce it from "Endangered" to "Threatened" and then delist the species'. This deceptively simple objective is then qualified by a list of the ambitious criteria needed to demonstrate recovery before this can occur, as follows.

This objective will be achieved when further loss, fragmentation, or degradation of suitable, occupied habitat within the butterfly's historical range in the Upper Florida keys and Miami-Dade County has been prevented; when the breeding sites of the Ssb have been protected from mosquito spraying; when mosquito spraying in the areas used by the Ssb has been reduced by 90 percent; when all suitable, occupied habitat on priority acquisition lists for the Ssb is protected either through land acquisition or cooperative agreements; when the hardwood hammocks that form the habitat for the Ssb are managed, restored, or rehabilitated on protected lands; and when stable populations of the Ssb are distributed throughout its historic range. These populations will be considered demographically stable when they exhibit a rate of increase (r) equal to or greater than 1.

In turn, each of these themes was followed by a number of recovery actions (USFWS 1999).

The above case exemplifies a rather frequent situation in which a stated aim of recovery is to downgrade the species to a lower category of threat, or remove the species from a list of threatened taxa. As above, criteria should be defined for any such step.

A clear perspective statement on the grounds for a main objective can be useful in indicating further the basis for such an objective, particularly to a non-scientific readership. This commonly includes many of the managers on whom subsequent action will devolve, and whose understanding is thereby critical. Thus, for Hine's emerald dragonfly (*Somatochlora hineana*), USFWS (2001) noted that 'The recovery criteria are based in the available information for Hine's emerald dragonfly and related odonate species and on basic principles of conservation biology. As additional information on the life history, ecology, population dynamics, and current status of this species becomes available, it may be necessary to revise these criteria'. Criteria are then specified for reclassification of the dragonfly from endangered to threatened, and for de-listing. As examples, the first criterion in each category reads that 'Each of the two Recovery Units contains a minimum of two populations, each composed of at least three subpopulations. Each subpopulation contains a minimum of 500 sexually mature adults for 10 consecutive years'. For de-listing, the threshold is increased to read '. . . a minimum of three populations . . .' in the above. Additional criteria for the dragonfly emphasise the need for site protection and management, such as watershed protection including 'up gradient watershed'.

Box 2.1 · *Interpretation problems: the two sexes of an insect may prefer different habitats. Hine's emerald dragonfly in North America*

Hine's emerald dragonfly (*Somatochlora hineana*) is listed as endangered under ESA, and considerable efforts have been made to clarify its ecology as a basis for designing conservation management. Surveys of the distribution and habitat relationships revealed considerable differences between the two sexes and emphasised that conservation must extend well beyond the wetland areas in which the dragonfly breeds, to incorporate also neighbouring dry meadows and other upland areas (Foster & Soluk 2006).

The two sexes show very different patterns of habitat occupation. Males predominantly stay in wetland areas, but females avoid these and are found mainly in dry meadows, forest clearings and similar places entirely unsuited for breeding. They return to wetlands in late adult life to mate and lay eggs.

Earlier, misleading, interpretations of sex ratio in this species had been based on samples from wetlands, but such sex-specific differences in habitat use counsel the need to carefully investigate any such possible sampling artefacts. Foster and Soluk found that food supply was not markedly different in the two habitat categories, and suspected that females may move away from water to avoid peak male activity and harassment whilst they are young.

A major lesson from this study, probably to be found in numerous other insects for which sex ratios appear to be biased, is that the two sexes are not always together but may differ significantly in some aspects of their behaviour and biology. In this case, apparent non-habitat areas near breeding sites may be a critical resource: dry upland areas are part of the complex habitat mosaic needed to conserve *S. hineana*, and parallels may usefully be explored for other dispersive insects, not least to ensure that such initially unlikely habitat components are not overlooked.

As a second example, in the plan for the endangered Mitchell's satyr butterfly (*Neonympha mitchellii mitchellii*), USFWS (1997) stated that reduction from endangered to threatened would occur when '16 geographically distinct viable populations or metapopulations are established or discovered range wide', with a stated allocation for those colonies distributed across the range, and 'at least 50% of those sites will be protected and managed' to maintain the butterfly. Delisting would occur only when an additional nine such populations (for total of 25) were established or discovered and remain viable for five years and a minimum of 15 sites must be protected and managed for *N. m. mitchellii*. Additional information clarified what is meant here by 'a site', as follows: a site for this fenland butterfly should have four components in order to qualify as a viable population (USFWS 1997).

1. A reasonable expectation of 300 individuals per brood, on average, for 5 of 7 years, with no fewer than 50 individuals in any given year, and a stable or increasing population.
2. A protected core of occupied habitat sufficiently large to allow for a mosaic of natural wetland vegetation types which are maintained by management or natural processes.
3. An adequate upland buffer of natural vegetation around the occupied core.
4. A landscape surrounding the occupied core that maintains the quality and quantity of the groundwater feeding the wetland.

Objectives should also be based on biological information to the greatest extent possible but, for the great majority of threatened insects specific information on population sizes and viability is not available for incorporation into this purpose (see Schultz & Hammond 2003 on a lycaenid, *Icaricia icarioides fenderi*, for application of such information). Nevertheless, information on the nature and intensity of threats and their abatement is at the core of formulating good management. Additional research on the biology of the focal species is almost always necessary to elucidate these, so that many recovery plans address the twin themes of 'research' and 'management'. It is all too easy for the research demands to become loosely focused in not specifying the precise information needed to enhance understanding for management. Again, specified timelines for both duties and review may be vital to ensure that such basic research work does not become indefinite and an end in itself, notwithstanding the values of continuing to accrue data on any threatened insect species.

Critical self-appraisal and discipline may be needed, and the flow scheme designed by Sherley (1998) gives useful pointers (Fig. 2.3). In short, initial review of all available knowledge of the target species, of its needs, and of the threats present may enable some initial appraisal of whether knowledge and understanding is sufficient to design effective management and move directly to this stage. Conversely, it may highlight areas in which fundamental knowledge is lacking, so that strongly focused management may be unwise. Any management undertaken may be *in situ*, *ex situ*, or both of these. Should more information be needed, this will usually be on aspects of population size and dispersion, life history, diet, habitat relationships, resource supply and/or threat evaluation. Constructive investigation of any of these topics may demand development of original methods or approaches. In this particular case (for weta), the recovery management group also had defined responsibilities, and similar terms of reference are a valuable component of any similar exercise, in relation to setting priorities, evaluating progress through monitoring and review, assembling and reporting new information, adjusting recommendations and, where necessary, helping to obtain the resources needed for the project to be carried out effectively.

Ensuing 'Actions' from objectives must also be very clearly formulated and, as for the embracing objectives themselves, should flow naturally from each objective and be accompanied by measurable criteria to enable monitoring. It is intriguing to contrast the plans with equivalent intent arising from the United States Endangered Species Act (US, as Recovery Plans) and the United Kingdom Biodiversity Action Plan (BAP, as Species Action Plans). Both suites of plans are the key references

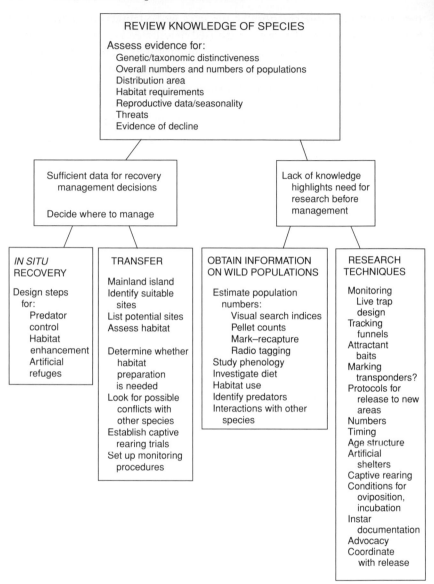

Fig. 2.3. A sequence for designing an insect conservation plan, and the practical decisions needed at various stages: this example is for weta in New Zealand (after Sherley 1998).

for conservation promotion and action in those regions. The disclaimer for US plans notes 'Recovery plans delineate reasonable actions which the best available science indicates are required to recover and/or conserve listed species', and plans 'are subject to modification as dictated by new findings, changes in species status, and the completion of recovery actions'. Details of recovery actions proposed in US plans are generally preceded by comprehensive summaries of the biology and conservation needs of the species, commonly occupying several tens of pages, or more. These plans may become lengthy; that for the Karner blue butterfly (p. 82) is 239 pages in length, for example. Need for research may be reflected in 'Recovery' by such means as designating 'interim criteria' for meeting objectives, pending further research, and clear statements of the need for additional work in order to evaluate threats fully: see, for example Tansy (2006, on a poorly known water beetle). Many objectives are local in application, such as for particular sites, and draw on knowledge of related species as appropriate. Many are also open-ended, and implementation schedules appear only irregularly. Objectives and actions are commonly accompanied by extensive 'step down' commentary, often including separate appraisals for the different sites from where the species is known. In contrast, the UK BAPs are typically very brief, of around two pages, with brief statements of biology, conservation status and needs. Proposed actions are simply listed, with lead agencies sometimes designated, but time lines and other aspects of 'SMART' may be difficult to discern. A series of UK Butterfly Action Plans are rather more fulsome in content. They are more comprehensive summaries of species biology and past conservation actions but, again, not committing most of the actions to any timing schedule. That for *Hesperia comma*, for example, contains these details for only two of 26 objectives ('Conduct surveys of all colonies and potential habitat every 5–10 years'; 'Review this Action Plan annually and up date in five years', elsewhere in the document specified as 'in 2000') (Barnett & Warren 1995). This same deadline persists in a more recent online version of this plan (accessed 30 March 2007). However, in the working climate for insect conservation in the United Kingdom, such brevity of treatment may not be a disadvantage. Awareness of conservation need is high, the interests of the numerous volunteer conservation groups and naturalists are not deterred by the formality of imposed action deadlines and reporting dates, and 'things get done'. Thus, the BAP for the hornet robberfly (*Asilus crabroniformis*) lists 18 more local British Action Plans specifying concerns for it; and that for the stag beetle (*Lucanus cervus*), a similar suite of 17 local plans, as well as various 'advice leaflets' containing much practical information for surveys and conservation measures. In such instances, more detail may indeed not be needed, notwithstanding the wider standard ideals implied earlier, and the central plan serves as an effective umbrella guide for others to elaborate and prosecute, often from considerable local knowledge, interest and expertise. Thus, surveys for *L. cervus* in Britain in 2002 involved some 1300 recorders (Smith 2003), a resource simply not available in most other parts of the world. Likewise, surveys of butterflies and larger moths in Britain can draw on the expertise and coordination of the organisation Butterfly Conservation, with a membership exceeding 12 000 interested people. UK species action plans range from promoting surveys to determining major practical recovery efforts (see Stewart & New 2007). For taxa, places and ecological contexts for which education is a more important and central consideration, more background information is needed. Undertakings from many US plans must be pursued in arenas in which biological information is limited, and commonly through the lead of government agencies with substantial other interests and priorities. The comprehensive leads provided by detailed recovery plans may then be invaluable.

Australia is in a somewhat intermediate position between Britain and the United States, with interest and capability for insect conservation starting to gain momentum, but considerable further impetus and education necessary. As elsewhere, and emulating examples in other parts of the world, insect recovery plans vary considerably in content and value, with no agreed national standards for these, and most designed at State/Territory level. That for the lycaenid *Hypochrysops piceatus* in Queensland (Lundie-Jenkins & Payne 2000) is a particularly valuable model, and contrasts markedly with the much less focused 'Action Statements' (these, however, designed with the lesser objective of being 'brief management plans') available for some listed insect species in Victoria. The hierarchy of general objective, specific objectives, recovery criteria and progress criteria for *H. piceatus* is clearly expressed, with the progress criteria linked firmly to specific objectives. Actions listed are precise, responsibility is defined, and all are budgeted appropriately. A useful aside from that plan is that it also notes wider biodiversity benefits, so broadening its appeal and relevance.

As Boersma *et al.* (2001) noted, the effectiveness of recovery plans (with the desired stated outcome being 'recovery', where possible accompanied by de-listing, and with possible continued conservation interest as 'rehabilitated species': New & Sands 2003, see p. 229) can usually be measured only by some form of trend analysis rather than by an absolute outcome. Measuring such trends depends on objective criteria. They suggested that this capability improved in plans that have been revised, with the important caveat that this betterment might reflect duration of attention rather than just new information. Some revised plans revealed new knowledge but not revised management recommendations. Such inferences from a critical review of US plans suggest that similar overviews elsewhere could be a valuable contribution to enhancing their value in assuring (or, at least, maximising chances of) recovery.

At present, there is little room for any widespread complacency over the design, content and prosecution of recovery plans and related documents for insects. With the limited expertise available for practical long term programmes for conservation of insects, any improvements we can foster are surely worthwhile in enhancing both the practice of insect species management, and its credibility. Focusing more clearly on well defined objectives while designing plans, ensuring their timely review and revision, and clearly integrating research and management components, appear to be highly rewarding aspects of such endeavour. It should be noted that objectives of management must focus on a particular level, normally a population (addressing single populations or metapopulations, usually by definition on a defined site or series of sites) or more widely, as taxon objectives with taxonomic integrity implicit in documents purporting to be 'species management plans'. Another categorisation of objectives differentiates between 'state objectives' (such as population size) and 'process objectives' (such as increase in population size, extension of distribution) for which the questions asked are of the form 'has x occurred?'.

Pavlik (1996) also distinguished between short-term (proximal) objectives and long-term (distal) objectives, to distinguish between desirable more immediate outcomes and longer term sustainability and ecological integration.

In practice, what sort of objectives and actions appear in plans? Most objectives drive toward site protection and prevention or amelioration of threat(s), as a rational basis for planning and underpinning management. It follows that understanding the variety and effects of possible threats is a central theme in species conservation. The threats faced by insects, in common with those to other biota, broadly encompass the conventional four categories of habitat loss or change, effects of invasive species, pollution and, to a rather less common extent, over-exploitation. Considerations of climate change, perhaps the most pervasive form of pollution to consider over coming decades, now override many of the more familiar threatening processes, with some insects amongst the most sensitive harbingers of the changes involved. However, whereas the categories of threat are very general to animals, insects manifest important issues of scale in assessing their impacts.

Consider two Australian examples:

1. The skipper butterfly *Ocybadistes knightorum* is known from about five small subpopulations in very small areas, spanning a distance of only a few kilometres of coastal subsaline peatland in central New South Wales (Australia) and, despite extensive searches, has not yet been found elsewhere. It thus appears to be a genuine narrow range endemic taxon, whose future depends entirely on the protection and management of the sites on which it occurs together with the sole, highly localised larval food plant, *Alexfloydia repens*. It has no economic impact and is unknown to most people other than lepidopterists. *O. knightorum* exemplifies that many insects thrive in very small areas, sometimes of a hectare or less, which would be disregarded and dismissed as unimportant for many other animals through being too small to support a viable population. Such tiny patches of habitat may be critical for many insects.

2. The Bogong moth, *Agrotis infusa*, undergoes long distance annual migrations to and from high altitude aestivation sites in the Australian alps. These migrations constitute a unique biological phenomenon in Australia, and cause media comment in most years when vast numbers of moths invade Parliament House, Canberra, or various sports arenas in early summer when attracted by lights. Indeed, an advisory briefing paper on *A. infusa* was produced recently by the federal Parliamentary Library (McCormick 2006), and stimulated by the moths entering Parliament House. Of rather wider relevance, as a threat to the image of this iconic species, *A. infusa* appears to be a vector of arsenate from lowland cropping areas to remote highland sites, with possible side effects by poisoning alpine endemic vertebrates such as mountain pygmy possums (*Burramys parvus*) that feed on the moths in its aggregation sites. Its conservation must therefore include security of alpine caves and rocky areas as specific aggregation areas, together with wider and less defined lowland breeding areas, as well as measures to reduce any perceptions that it is itself a threat to more charismatic animals. Rather than just focusing on single sites alone, conservation here moves to a landscape or 'area-wide' perspective.

This combination of scales is often an important consideration. The monarch butterfly (*Danaus plexippus*) occurs throughout most of North America and thus has a continental distribution. However, its survival depends on the conservation of very limited forest overwintering sites in California and Mexico (Brower 1996). Adults aggregate there in vast numbers, and without protection of those forests from logging and other despoliation, the entire North American population of *D. plexippus* is at risk.

Summary

1. Sound 'recovery plans' or 'management plans' form the foundation of insect species conservation, and must be constructed logically and based on the best possible scientific knowledge. Provision is necessary for monitoring, and for management to respond to changes that occur in the species' population. Plans incorporate both regulatory and scientific aspects of conservation, and may be used to prevent further losses or to more actively promote recovery to higher numbers or wider distributions than occur at present.

2. Most such plans are for individual insect taxa, but others are for groups of species or higher taxa. They can be made at a variety of scales ranging from local to global. Broader plans can incorporate generality and may provide useful practical focus for more local application or for individual included species. Many conservation plans flow from legal obligations resulting from 'listing' the species for formal protection.

3. Although there is no universal format for an insect recovery plan, one prime consideration is that much of the readership (and people responsible for bringing it into practice) will have little practical experience with, or knowledge of, insects. To this end, a plan must contain sufficient information for effective communication, with an ordered sequence such as (a) expressing the case for conservation, with background information on the biology of the species, (b) summarising actual and likely threats, (c) developing remedial measures, (d) the management actions and details needed to control threats, and (e) how these may be undertaken and monitored. Each of these may require additional research to be satisfactory, but all previously available information should be gathered and appraised.

4. The objectives (aims) of a plan should be specified very clearly and, with ensuing actions, expressed in SMART terms. Where possible they should also be costed. The amount of biological background

given in any plan may depend on local policy requirements and on the potential readership. Criteria for 'recovery' or success of management should be specified realistically, and both short- and long-term objectives specified.

5. All possible management options (including 'doing nothing') should be evaluated.

3 · *Habitat, population and dispersal issues*

Introduction: concepts of habitat

To an insect, the world consists of a hierarchy of habitats, which many ecologists divide somewhat arbitrarily into 'macrohabitats' and 'microhabitats'. Hanski (2005) used the felicitous term 'habitat matrioschkas' to reflect this hierarchy of scales, whereby habitats are nested in the same manner as the famous Russian dolls. He exemplified the concept by referring to the European saprophilous pythid beetle *Pytho kolwensis*, for which the relevant matrioschka has the sequence: boreal forest; spruce-dominated forest; spruce-mire forest with high temporal continuity of fallen logs; a fallen spruce log with the base above the ground; a particular stage in the decay succession of phloem under the detaching bark. The first three of these were regarded as macrohabitat, and the last two as microhabitat. Reflecting their small size and ecological specialisations, many insects depend on resource-based 'microhabitats' for their well-being and sustainability, but these in turn depend on the continued presence of the embracing macrohabitats. Thus, attributes of 'place' (categorised broadly by major ecosystem: here, boreal forest, and commonly referred to as 'biotopes') intergrade with more specific needs that may be viewed as progressively more tangible resources at finer scales. The major practical lesson, as emphasised by Hanski (2005), is that much of insect species conservation planning must heed and focus on microhabitats, commonly to a far greater extent than for many vertebrate conservation plans. Whereas many a bird or endemic marsupial in Australia may have its habitat classified satisfactorily merely as 'eucalypt woodland' or some similar broad descriptive term, most insects found in that vegetation association will have far more precise needs and defining their habitat will need correspondingly finer descriptors, as with *Pytho* in southwestern Finland. Many insects, for example, will depend on particular plant species and some, on particular plant stages, structures and condition, as well as on other insects present, to provide very precise requirements for food and reproduction. Many insect microhabitats may be short-lived and need intensive management to sustain them or to assure their continual supply. It is worth noting here the frequent confusion or implied synonymy between 'habitat' (as place of occurrence, without other implication) and 'biotope' (as the area with characteristic associations of plants and animals as the biotic association within which the species occurs). The second of these leads to consideration of a species' ecological specificity and specialisations.

Resource-based definitions and practical concepts of 'habitat' are thus particularly important in insect conservation. Rather than just 'a place to live', a habitat is also a place where a suite of critical resources needed by a particular insect species coincides and is sustained (see Dennis *et al.* 2006). This requirement may be much more complex than initially evident. First, many insects have markedly different requirements at different stages of their life cycle, and the requirements of all the various life stages must be present or accessible. As examples, (1) many insects have aquatic larvae and terrestrial adults, and (2) the majority of insects undergo a complete metamorphosis wherein larvae and adults differ greatly in appearance and requirements. Thus butterflies and moths typically have larvae, caterpillars, that are chewing herbivores, often restricted to particular plant species or stages for sustenance, and adults that feed on angiosperm nectar, and which may respond only to particular chemical and visual cues for oviposition. Additional ecological 'complications' are not unusual. Many lycaenid butterflies of great conservation interest have caterpillars that form obligatory mutualistic associations with particular ant species, so that the minimum obligate needs of the species comprise three very diffent categories of resources: availability of a particular ant species, a particular larval food plant, and nectar sources for adult butterflies. Such combinations of need ensure that a suitable habitat is far more than just a place that can be designated reliably by a fence or legal instrument. The particular spectrum of resources needed varies enormously between species, and in level of specialisation. Thus caterpillars of the Karner blue butterfly in New York may be attended by 19 species of ant (in three subfamilies), some much more commonly than others (Savignano 1994), whereas those of many Lycaenidae form mutualisms with single ant species, with even their closest congeners unsuitable substitutes.

However, many legal documents rely on the 'place' emphasis in definitions of habitat for their undertakings, in some instances as a legacy of vertebrate-influenced origins. Thus, the important concept of 'critical habitat' under ESA is defined as '(1) specific areas within the geographical area occupied by the species at the time of listing, if they

Box 3.1 · *Intricacies of synchronising an insect with a critical resource: seasonal life cycle patterns of the Karner blue butterfly (*Lycaeides melissa samuelis*) and its sole larval food plant, wild lupin (*Lupinus perennis*), at a site in New York*

A survey by Dirig (1994) emphasised the importance of understanding the natural history of both the insect and the sole food plant, in an example of wide relevance for any insect of conservation concern. The seasonal patterns of development of the monophagous Karner blue butterfly and its larval food plant (Fig. 3.1, p. 90) demonstrate the key stages for both species, and how the insect development is linked with this critical resource. Details of both phenologies are likely to vary geographically, so that any particular site-specific study may only provide a general guide to management need of a species and its resource. The pattern of resource availability may be a critical determinant of the insect's wellbeing.

At Dirig's study site in Albany, lupins sprout from perennial subterranean rhizomes in April, and plants grow rapidly to bloom fully by the last third of May, leading to development of seed pods by mid-June. The pods dehisce to spread seed around 1–2 m from the parent plant, and these may germinate rapidly or in the following season. Seedlings die off with autumn frosts.

The Karner blue overwinters in the egg stage, and hatchling caterpillars feed on young lupin foliage. First generation adults fly from late May to mid-June, with offspring from these developing to produce second generation butterflies in late July–August. Their overwintering eggs complete the bivoltine cycle of development.

contain physical or biological features essential for conservation; and (2) specific areas outside the geographical area occupied by the species if the agency determines that the area itself is essential for conservation'. Designation of such critical habitat is a condition of listing any species under ESA, and must be based on the best scientific information available (within provisos of an open public process and specific time period). An obligation to United States federal agencies in protecting a listed species following such designation is then not to 'destroy or adversely modify its designated critical habitat' (section 7, ESA). Similar approaches occur elsewhere. In New South Wales, Australia, the State agency has provision to identify and declare critical habitat in protecting a species listed as endangered, where this is defined as 'areas of land that are crucial to the survival of particular threatened species, populations and ecological communities'. The term thus refers directly to 'area' and is reiterated as such in the ensuing declaration process. Species management must then include an impact statement for all developments and activities that affect declared critical habitat. However, flexibility occurs: the recovery team for the Bathurst copper butterfly (*Paralucia spinifera*), for example, determined that adequate habitat protection is available without such designation, and that other avenues were likely to be more effective in its conservation (NSWNPWS 2001).

Concepts of habitat designation based on 'space' including the resources critical to a species' survival – with use of terms such as 'patch' and 'matrix' – are implicit in many current perceptions and, indeed, in elucidating population structure of many insects of conservation interest. The important idea of metapopulations, for example, was developed largely from studies on distribution of butterflies across landscapes, and marks a considerable perception change from the formerly predominant ideas assuming that these insects occurred generally in closed (discrete) populations. It has led also to need to re-interpret the significance of local extinctions, rather than presuming that all such disappearances are catastrophic – and, hence, of some patterns and needs for proper conservation management. As emphasised in several recent papers by Dennis and his colleagues (see Dennis *et al.* 2006), an alternative approach is to view landscapes for insects as continua of overlapping resource distributions. For metapopulations, there may be no single condition for an optimal habitat, because the entity flourishes across patches of widely varied suitability and which vary continually in that suitability. Carrying capacity of any given site for a given insect species, and even its capacity to support a resident population, will vary over time, and reflect either cyclic or directional changes in resource availability.

For each insect species, or life stage of that insect, the habitat may be viewed as two interacting and complementary suites of resources, which Dennis *et al.* (2006) termed 'consumables' (such as host plants, hosts, nectar resources) and 'utilities' (physical sites for mating, pupation, hibernation; and suitable conditions for development such as microclimate and enemy-free space). The second group, although intuitively obvious, has often been ignored in insect conservation programmes, and many management plans have focused largely on 'consumables' alone, as these are the most obvious and easily defined needs of a species. Likewise, many plans have not adequately acknowledged the effects of weather or season on the species involved. Thus, in the golden sun-moth noted earlier, weather conditions (low wind speed, no precipitation, adequate temperature) are critical variables affecting detectability. Should unsuitable weather prevail over the few days of life of individual male moths, it is likely (though not proven in the field) that they may not mate because they would not then fly and detect (or be detected by) non-dispersive receptive female moths. Meeting of the sexes in this species occurs through the females exposing their brightly coloured hind wings in response to males flying overhead, a metre or so above the short grasses used by the moths and on which the subterranean caterpillars feed. Once 'signalled', the male lands and mating ensues. In such species, a long flight season may counter such short-term weather vagaries – perhaps paralleling the inter-seasonal variations countered by variable diapause, noted earlier (p. 30). Interpreting temporal variability is an integral part of the understanding needed to optimise insect conservation management, not least because 'apparency' is critical in monitoring the progress of any management measure proposed or undertaken. Some workers (such as Heikkinen *et al.* 2005) delimit three suites of variables, rather than two, namely 'habitat', 'resources' and 'microclimate'.

Habitat models

Whatever such categorisation is employed, for the great majority of insects there is a severe dearth of information of sufficient detail from which to predict the features of an ideal habitat or wider environment. Both consumables and utilities are commonly not known sufficiently for prediction of 'good' habitat on any quantitative basis. Most evaluations of insect habitat 'quality' or 'suitability' have resulted from correlation between a few selected features (usually of 'consumables') and numbers (occasionally simply presence) of the insect of concern. As one way to help overcome this, at least in part, several workers have recently explored the use of so-called 'habitat models' for species conservation, as a means to investigate patterns of habitat use by correlating presence of the insect with a wider array of environmental variables. Although still not 'proving' causation, this approach may be especially helpful for species for which broad habitat restrictions can be defined easily – such as 'salt marsh' or 'grassland' – and for which the relevant parameters can thereby be focused somewhat to the variables evident wthin that system, and for species that are believed to normally disperse rather little so that populations are faithful to circumscribed sites. Correlation of features with abundance of the insect may then reveal the 'most desirable' factors for incorporation in management. Such models thereby (1) analyse the relationship between a given insect species incidence and abundance in relation to environmental features and (2) predict its incidence and abundance given certain environmental conditions. Thus, for grassland leafhoppers, Strauss and Biedermann (2005) used this approach to construct 'habitat suitability maps' that facilitate prediction of habitat quality under different management regimes. The leafhopper *Verdanus bensoni* is a rare species of montane grasslands in Germany, and two parameters were sufficient to reveal a high habitat suitability for this species. The most important habitat factor was fertility of the grassland sites. Because *Verdanus* is limited to low productivity sites, agricultural intensification and fertilisation of these sites poses a threat. Aspects of land management for agriculture are then a key conservation consideration. The occurrence of *V. bensoni* would also decline if tree cover increased, so that abandonment of mowing or grazing might also pose a threat as succession proceeds (Strauss & Biedermann 2005).

The approach may be used on different geographical scales, to assess the relative importance of the variables employed. For two endangered Lepidoptera, Binzenhöfer *et al.* (2005) developed separate habitat models based on (1) whole sets of available data and (2) parameters available for the whole study area, so that the latter, wider, approach allowed construction of area-wide habitat suitability maps using logistic regression analysis. The two species studied were both associated with semi-natural dry grasslands, but primarily with rather different successional stages within this declining habitat type. The burnet moth *Zygaena carniolica* is associated with moderately grazed or mown dry grasslands, and the butterfly *Coenonympha arcania* inhabits dry grassland with bushes. Biological differences include larval food plants and adult nectar resources. The moth prefers violet-flowered nectar plants, for example. The presence of suitable nectar plants (with two key species named) and management type (highest occurrence associated with sheep herding and lowest with cattle grazing and mulching) were strong correlates with *Z. carniolica* abundance. *C. arcania* was correlated strongly with extensively mulched grasslands with nearby bushes and hedges.

An example for Coleoptera involves the endangered European ground beetle *Carabus variolosus*, a riparian species in southern Europe (Matern *et al.* 2007). Habitat mapping showed that it is restricted to fringes of water bodies and to areas with high soil moisture and with patches of bare soil. It avoids areas with acid soil and slightly favours those with sparse trees.

Box 3.2 · *Predicting habitat suitability for conservation: Simson's stag beetle in Tasmania*

A flightless endemic Tasmanian lucanid, Simson's stag beetle (*Hoplogonus simsoni*), was the subject of a predictive study based on geographic information systems (GIS) with a variety of habitat variables examined to extrapolate the features of its known habitat to estimate the additional habitat area potentially available to the species. Meggs *et al.* (2003, 2004) found the optimal habitat for this threatened beetle to be wet eucalypt forest below 300 m altitude, with slope <5°, deep leaf litter and a well-developed shrub layer. Deep, well-drained soils were associated with high beetle densities.

    H. simsoni is distributed very patchily over a range of about 250 km^2 of northeastern Tasmania, and Meggs *et al.* (2003) used data from five major forest types in the region (with a total of 42 locations and 252 sites, the latter each with 6 plots of 1 m × 1 m, to assess fine scale microclimates). Three predictive models were applied, showing respectively that (1) *H. simsoni* decreased with increasing altitude but that this parameter was a poor predictor of abundance; (2) greater frequency occurred at sites with southern and eastern aspects; and (3) there is some overall potential to predict abundance.

    This study enabled Meggs *et al.* to identify the extent and distribution of suitable habitats for *H. simsoni* in forest areas in which forestry activities are important, so leading to potential harmonising of effective conservation of the beetle within a significant industrial context, despite absence of detailed information on many aspects of the biology and ecology of the beetle.

The management recommendations from this study included banning forestry from riversides and beetle habitats, to maintain water levels. They also stressed the importance of floodplain and headwater areas as habitats for *C. variolosus*.

    In a somewhat different context, translocations or re-introductions (p. 183), habitat models can be an important tool for helping to predict site suitability and so to select and rank sites for the exercise. Such approaches incorporate considerations of both the insect and its critical resources, particularly plants. One example is for the lycaenid *Cupido minimus* in north Wales, for which the dynamics of the larval food plant were a critical consideration (Leon-Cortes *et al.* 2003b). Data on relationships between the butterfly and its sole food plant (caterpillars feed only on inflorescences of kidney vetch, *Anthyllis vulneraria*) from extant butterfly colonies were used to generate models of habitat suitability over a wider historically occupied area, and to assess their possible suitability for re-introduction success. Kidney vetch had declined considerably over north Wales, reflecting loss of limestone and sand dune habitats, but considerable areas of apparently suitable habitat remain. The vetch varies considerably in abundance from year to year, with these dynamics greatly affecting the outcomes of re-introduction simulations. Modelled patch occupancy by *C. minimus* was affected greatly by this, with the butterfly usually becoming extinct when vetch density became low.

More generally, Matern *et al.* (2007) summarised the possible uses of habitat models in insect conservation as:

1. Assessing the relative importance of habitat variables to a species.
2. Determining differences in the habitat use by a species, even on a small scale.
3. Assessing the habitat quality of surroundings of populations of focal species.
4. Predicting the effect on species occurrence of environmental changes within a habitat.
5. In conjunction with geographical data, helping in the development of conservation areas and management practices.

Superimposing management effects or templates of finer correlation across a range of sites on which the target species occurs may give valuable information on (1) which sites or populations may be most secure in terms of juxtaposition with critical resources and environment; (2) which aspects of any particular occupied site may require priority attention and what that attention might be, and (3) which of several available sites might be most amenable to restoration as a possible focus for translocation or re-introduction (p. 184). Using any such features to define habitat in such a way, mapping suitable habitat for an insect may commonly reveal that optimal habitat is much smaller in extent than initially thought, with populations having a collective area of occupancy far smaller than the apparent habitat available for them. The reasons for such restriction reflect aspects of the 'utilities' category of resources, as above. The Adonis blue butterfly, *Lysandra bellargus*, in southern England, is there on the northern fringe of its European range. At the time of Thomas's (1983) study, it could thrive only on south-facing slopes where insolation was sufficient to maintain warmth. Indeed, Thomas (1983) suggested the management need to increase availability of such aspect by using bulldozers to create additional suitable slopes – a measure likely to horrify uninformed onlookers! Importantly, though, this distribution indicated that aspect of a microsite may be an important component of an insect's thermal regime. Returning, again, to the golden sun-moth, surveys at different times of the season sometimes show that 'hotspots' of incidence shift across sites as summer progresses. This may reflect different levels of insolation or aspect influencing duration of development or timing of emergence; as New *et al.* (2007) put it 'hotspots may simply be hot spots', rather than reflecting distribution of consumables over a site.

Microclimate may, of course, affect availability of consumables or other resources. Grazing regimes have repeatedly been proved to be a critical aspect of management for chalk grassland butterflies in Britain, through which sward height directly influences ground surface temperature and the suitability of the site for the obligatory mutualistic ants on which the butterflies depend. Particularly when a number of species of interest co-occur, maintenance of different sward heights through different grazing regimes or 'grazing mosaics' in time and space may be needed (BUTT 1986). Such 'microtopographic effects' may be very subtle. The Karner blue butterfly (*Lycaeides melissa samuelis*) is one of the more intensively studied North American Lycaenidae. Grundel *et al.* (1998) examined the behaviour of this subspecies under varying degrees of canopy cover, because increased canopy cover was implicated as a major factor causing its decline in many places. Males used habitat under canopy much more than females, and oviposition was highest under regimes of 30%–60% canopy cover, even though the larval food plant (the wild lupin, *Lupinus perennis*) was more abundant in more open areas. However, caterpillars preferred to feed on larger lupin plants and those in denser patches, and shaded lupins were generally the larger, so that the extent of shade in part mediated the 'balance' between lupin abundance and distribution of oviposition and feeding. As with many other specialist insects, the timing of the various life history stages is linked very closely with the seasonality of the food plant (Fig. 3.1).

Need for initial surveys to examine distribution of threatened insects is an almost universal requirement in conservation programmes and, for many such insects, the particular places in a landscape that are likely to reward such effort are definable (sometimes only tentatively) on features of vegetation and/or topography. However, those apparently suitable

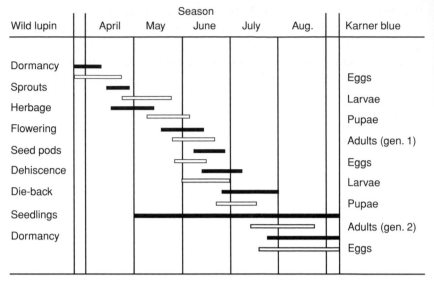

Fig. 3.1. The yearly development profiles of a herbivorous insect and its specific food plant; the developmental timing of the Karner blue butterfly (*Lycaeides melissa samuelis*) and wild lupin (*Lupinus perennis*) in New York (after Dirig 1994).

areas may be only very small patches within otherwise unsuitable but varied landscapes. They can easily be overlooked, with the outcomes that numbers of populations and overall abundance of insect species may be underestimated. Mawdsley (2008) explored the value of high-resolution satellite imagery in searching for patches suitable for threatened tiger beetles (*Cicindela* spp.). Following from more widespread earlier uses of topographic maps and aerial photographs to help locate suitable areas, Mawdsley used two World Wide Web-based tools (Google Earth, Microsoft Terraserver) to select patches for ground surveys, and believed such tools to have considerable potential for enhancing the efficiency and effectiveness of such surveys. One caution is that some of the resources needed by particular tiger beetles (soil salinity, for example) may not be detectable through the satellite systems presently available.

Species, resources and population structure in management

Much of the above discussion relates to the under-appreciated fact that site (patch) quality and extent are partners in conservation management for insects, most commonly at far finer levels than used conventionally for many vertebrate programmes. Dennis *et al.* (2006) distinguished four major groups of resource components that affect site or patch quality, namely composition (occurrence, density, multiplicity, variation and context), physiognomy ('geography': location, altitude, orientation, stage, area, height, slope, contagion, fragmentation), connectivity (contact and isolation, reflecting adult and larval mobility), and temporal aspects. We are thus concerned with the effective environment of a given species at the levels of individual site (harbouring a population or one or more metapopulation segregates) and the attributes of that site within the wider landscape actually or potentially available to the species. If a population is truly closed (so that immigration and emigration do not contribute significantly to its size, and numerical changes predominantly reflect births and deaths) and thereby also isolated from other populations of the species, the individual inhabited patch or site is obviously the primary focus for conservation management. With a metapopulation emphasis, though, largely independent demographic units manifest on different patches within a wider area, but with the entire metapopulation maintained through rolling 'extinction–recolonisation cycles' over a series of patches. Within this scenario, any individual extirpation of a metapopulation segregate may be entirely normal and, thus, of little long-term conservation concern. Without this knowledge, and particularly if a closed population structure is presumed, a local extinction or severe decline in abundance may be assessed as a crisis and trigger intensive measures to counter it, in situations where this reaction is not needed.

Box 3.3 · *The concept of a 'metapopulation'*

Many insects are purported to show a metapopulation structure, and this concept has been advanced considerably by studies on butterflies of conservation concern. Hanski and Gilpin (1991) defined a metapopulation as 'a set of local populations which interact via individuals moving between local populations'. This translates to a system of largely independent demographic units, each on a patch of habitat within a landscape and in which those patches are scattered in a non-habitable 'matrix'. The occupied patches thus each correspond to a putatively isolated site on which the insect may become a conservation focus. However, if the population is one of a number of 'metapopulation units' collectively found on a number of such patches, it is functionally

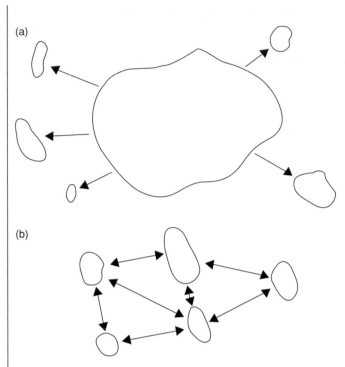

Fig. 3.3.1. Two major categories of metapopulation. (a) The 'continent–island' model, whereby a large, secure 'continental' population supplies individuals to smaller 'islands', whose populations are maintained by such immigration, and not all of which may be occupied at any given time; (b) a model representing continuing irregular interchange of individuals between demographic units (commonly 'sites'), each of which may or may not support a population at any given time.

linked with them by dispersal, and the whole metapopulation is maintained by a series of rolling local extinctions and recolonisations. At any time some occupiable patches are vacant, and local extinctions are entirely normal, rather than the subject of concern equivalent to that for a truly closed population.

Many kinds of metapopulation have been defined, but two broad categories are shown in Fig. 3.3.1. Figure 3.3.1a shows a 'mainland–island' system, whereby a largely secure 'reservoir' population on a large 'mainland' habitat patch provides a continuing source of individuals that colonise smaller 'island' patches. These can colonise patches on which prior extinctions have occurred (perhaps some years or generations earlier) or augment those populations. A second category

(Fig. 3.3.1b) involves the possibility of movements of individuals between habitat patches, as an 'archipelago' within the matrix. Again, not all patches may be occupied at any one time, but the presence of those patches within the dispersal range of the insect incorporates them firmly within the species' conservation consideration.

Each metapopulation unit is largely independent demographically for much of its existence but the pattern of extinction and re-colonisation ensures that the units are also interdependent. Regional persistence of the species requires that the rate of colonisation equals or exceeds that of local extinctions.

Harrison's (1994) essay provides considerable additional useful background to 'how metapopulations work'. She noted, for example, that increased habitat fragmentation can indeed transfer functional metapopulations to a series of isolated closed populations, so that studies of dispersal and functional connectivity become of critical importance in evaluating the real population state (and, hence, the level of conservation concern) of any insect species.

In practice, determining whether a metapopulation occurs is of fundamental importance in dictating limits to a conservation plan. Hanski's (1999) defined characteristics of a metapopulation have guided most later thoughts on how such entities 'work'. These characteristics are:

1. Breeding populations occupy discrete habitat patches within an area or landscape.
2. The habitat patches are sufficiently close to enable colonisation or recolonisation of empty ones.
3. The independence of each population (on a patch) is sufficient that not all of them will become extinct at the same time.
4. All the populations are individually at risk of stochastic extinction.
5. A substantial fraction of suitable habitat patches are unoccupied at any one time.

In practice, failure to recognise a metapopulation structure can confuse attempts to characterise 'good' sites on the basis of occupation, because the resources on occupied and unoccupied sites may differ little, if at all. The practical dilemma is distinguishing these population forms in real life situations and maintaining the landscape in condition sufficient to allow recolonisations to occur within the metapopulation, through preventing undue isolation of each potentially occupiable site, or by imposing barriers to movement between them. In a metapopulation, the individual population units are more or less distinct but connected by migration and gene flow, and the whole structure persists in a stochastic balance between local extinctions and recolonisations (Hanski 2005). Temporary absence may indeed reflect shortage of resources, and recolonisation indicate their recovery. Knowledge of any such fluctuations may inform the need for conservation, particularly in differentiating such 'blips' from longer-lasting successional change, which may indeed be associated with more permanent loss of insects as the site becomes untenable. Countering successional changes is a major, sometimes the only, feature of habitat management needed for many insects.

Box 3.4 · *Metapopulation theory to conservation practicality:* Maculinea *butterflies in Europe*

Metapopulation dynamics involve spatial fluctuations in incidence and abundance, but relatively few studies have examined how particular insects conform to a classic metapopulation theory, despite this often being inferred. This was one of two objectives of a comparative study of three species of *Maculinea* butterfly in Poland (Nowicki *et al.* 2007). The second objective was to identify what factors affect the butterflies' occurrence and abundance patterns, as of fundamental importance in studying their dynamics in practical conservation.

The study site, 35 km² of wet meadowlands, included 61 patches of *Sanguisorba officinalis* (the larval food plant of *M. teleius* and *M. nausithous*, mostly 100–300 m apart) and 18 patches of *Gentiana pneumonanthe* (food plant of *M. alcon*, mainly more isolated and separated typically by 300–700 m). Potential host ants (*Myrmica scabrinodis*, *M. rubra* and *M. ruginodis*) are all widespread in the area, but with densities varying considerably.

All three butterflies were found on almost all food plant patches, with the few vacant patches being significantly smaller and more isolated than the others. Patch size and shape were the most important factors affecting densities of *M. nausithous* and *M. teleius* (both found at highest densities on small and highly internally fragmented patches with higher ant densities), whereas *M. alcon* was limited more by food plant density.

Food plant patches and their surroundings are both important considerations for *Maculinea* conservation. The surroundings should be natural or semi-natural to be suitable for the ants, and elongated patch boundaries with internal gaps in food plant cover may be beneficial. In such case, the more usually suggested practice of planting additional food plants in such gaps might, in fact, decrease carrying capacity, rather than increase it. Nowicki *et al.* also noted the possible regulatory novelty of needing to conserve patch surroundings, as areas not directly inhabited by the focal species.

Victoria's only endemic butterfly species, the golden-rayed blue, *Candalides noelkeri*, is known only from two small sites in the Western District of the state, where larvae feed on a single host plant species, *Myoporum parvifolium*. Both sites are being invaded by the native *Melaleuca halmaturorum*, which poses a serious threat to *Myoporum*, with the density of *Myoporum* correlated negatively with that of *Melaleuca*. As well as direct losses of *Myoporum*, the quality and attractiveness of the remaining plants is declining through being shaded out. Because *Melaleuca* is a native plant species, it is less obviously intrusive to the uninformed observer than an exotic invasive, but this example demonstrates the attention that must be paid to succession and regeneration in conservation of the many ecologically specialised insects confined to such equally restricted and specialised environments.

In order to determine population structure for an insect, information on dispersal is among the most useful that can be acquired. Parameters such as distance normally moved, differences between the sexes, extent of movement between given sites, and which landscape features may constitute barriers or facilitate dispersal (such as by corridors) may all be relevant in designing management on a landscape or more local level. The most usual tool for this is Mark–Release–Recapture (p. 198), wherein captured insects are given individual or batch marks or individual numbers, released unharmed, and their fate assessed by future captures. Monitoring and assessment of population size by using such techniques may contribute substantially to ecological understanding. However, the numbers of individuals in populations of many of the species of concern are likely to be very low, so that this approach may not be feasible. In addition, risks of damage to small and delicate insects by capture and marking need to be considered carefully, and may preclude this approach for some species.

The importance of discovering which resources are indeed the major determinants of how an insect species is distributed is commonly not appreciated sufficiently in practice. Often, far greater emphasis is placed on assuring site security for places where the focal species occurs, and using the features of that site (or set of sites) as a standard to be sought or copied elsewhere. This approach is, on the face of it, entirely logical as those sites are the only examples of 'suitable habitat' that we have. However, it is almost always unclear whether the quality of those sites is indeed optimal. In some cases we may be endeavouring to restore or replicate habitats that are little more than marginal in quality and which may offer only limited carrying capacity – simply because these are the only models to hand.

The resources needed by a particular insect may not always be obvious. The idea of 'territory' as a critical resource for many insects, for example, may not be appreciated without knowledge of the particular insect's behaviour and mating strategy. Many male butterflies and dragonflies, probably in common with many other insects whose behaviour is less well known, use territories as perches or patrol areas for mate location. They may need very specific perching sites to which individuals return repeatedly, or more extensive areas (such as alongside forest edges or streams) for more extensive patrolling. The former may need to be elevated, exposed to particular microclimates (such as shade or sunlight), and allow visibility over the nearby area, and detailed needs may vary greatly for different species. For the British chequered skipper butterfly (*Carterocephalus palaemon*), Ravenscroft (1992, 1995) noted the common features of patches selected for territories as (1) relatively open in relation to nearby vegetation and with a few well-spaced perches; (2) specific heights of perches varying but selected according to visibility and temperature; and (3) sheltered, hot and often close to areas where female butterflies gather to feed on nectar. This species further exemplifies the numerous species of insect in which the two sexes may behave rather differently, and even frequent different habitats at times. Even recording the sex ratio in the field may be biased without knowledge of such differences. For the golden sun-moth (p. 30) males are active, but females almost wholly flightless, cryptic and difficult to find. In such cases, as with the many insects in which one sex is flightless (or much less dispersive, or more cryptic than the other), inferences of different behaviours are clearcut, but many others are not as evident without close observation. Female *C. palaemon* spend much of their time in herb-rich areas with abundant nectar plants. Males disperse to territorial sites with rather different characteristics. Further, because the females may disperse widely, the sites on which adults are recorded are not always suitable for breeding, because adult and larval resources may not occur together, adding a further complication to definition of 'critical habitat'.

On an even wider scale, the phenomenon of hilltopping is widespread among butterflies and many other insects. Individuals of species normally distributed widely across a landscape gather on elevated points – typically hilltops, as the name implies, although features such as isolated trees or other prominences in the landscape may be used – where access to mates for such widely dispersed taxa may be increased. There, males of territorial species may perch or patrol, and approach females as they reach the site. After mating, the females are presumed to move back to breeding areas. For some species, the whereabouts of those breeding areas is unknown, but may be up to at least several kilometres away. Presence on a hilltop provides evidence of the occurrence of the species in the region (and is a useful monitoring tool), but may not tell us much about its breeding requirements. In at least some cases, such as at Mount Piper, Victoria (Britton et al. 1995), these assembly sites are not breeding sites for some significant species found. The hilltops are nevertheless critical to the insect's normal behaviour, and it may be important to maintain vegetational complexity, with a variety of territory attributes, there. Indeed, clearing of vegetation from hilltops (for example for building telegraphic relay or repeater stations) is a declared threatening process for insects in New South Wales, because of this concern.

Still further complications arise with the realisation that not only may the resources needed by a given insect vary across its range, but the biology of the insect itself may differ considerably in different places. A study of the same insect species in different parts of its range may give us very different inferences for management, with differences in seasonality, food preferences and environmental tolerances of many kinds. Although differences in flight season may be the most visible of these differences to a casual observer (and may give characteristic patterns, so that the golden sun-moth consistently flies a few weeks earlier in inland Victorian sites than in those nearer the coast), the number of generations may differ, the preferred larval host plant may differ, and the main mutualistic ant species of Lycaenidae may change, among many other labile features. Thus, caterpillars of the Eltham copper (p. 208) in Victoria associate with different species of *Notoncus* ants in different places (Braby et al. 1999), and caterpillars of the Duke of Burgundy (*Hamearis lucina*) in central Europe may use different species of *Primula* for food on different sites (Anthes et al. 2008). Even within a site, individual plants and other resources may differ substantially in quality and attractiveness, sometimes as a consequence of other environmental features. Whereas *H. lucina* larvae, above, are monophagous on *Primula*, female butterflies may ignore plants under closed forest canopy for oviposition, so that part of the potential caterpillar food supply is not available to them, even though we may assume it will be used, and perhaps plan management on its availability.

Anthes *et al.* (2008) emphasised the importance of investigating such 'realised niches' along gradients in optimising management for larval habitat of *H. lucina*. Thermal regimes are also important for many other species (Davies *et al.* 2006).

The approach initiated by Dennis and Shreeve (1996; see also Dennis *et al.* 2003) merits adoption widely as providing biological knowledge of the suite of needs for each life stage of the insect to be conserved, and mapping these resources within the species' range. Thus, the minimum requirements for an adult butterfly would include resources for oviposition, mate location, resting, roosting, feeding and avoiding predators. Likewise, those for a caterpillar would include refuges for passing any period of diapause, pupation, feeding, avoiding parasitoids and predators, fostering any mutualism with ants and so on (see Dennis *et al.* 2006). Threshold values of any such resources are usually difficult to determine, but may be a basis for habitat augmentation and resource supplementation as a major facet of species management. For many species, the desirable parallel of increasing habitat extent is also necessary to counter continuing losses, but area may not always be available for this, either to increase the size of occupied patches or to found additional populations. For the latter strategy, translocations or re-introductions following site rehabilitation may be feasible. The latter may depend on captive-reared stock if sufficiently strong wild 'donor' populations are not available (p. 182). For the Lord Howe Island stick insect, *Dryococelus australis*, it is anticipated that stock reared from a pair of individuals captured on Balls Pyramid will eventually be released onto rat-free islets in the area (Priddel *et al.* 2002).

Landscape features

Consideration of metapopulations as a normal way of life leads firmly to wider, landscape level considerations of linkages between individual habitat patches, with the most obvious need being for linkages between those patches that are occupied at the time of a survey and, particularly, when these appear to be isolated increasingly as a result of fragmentation. Should dispersal activity (or, more generally, any form of ability to reach other habitat patches) be threatened by alienation of the intermediate terrain, the natural system of dispersal and colonisation inevitably breaks down. Metapopulations may not then be sustainable. Initial disturbance to a landscape may initially increase the number of patches available to an insect, simply because of division of the then existing large patches. Subsequent trends are (1) reduction in size of those fragments and (2) eventual loss of fragments, so that discontinuity is increased, and commonly associated with reduced habitat quality from loss of interior habitat and markedly increased edge effects. Many insects can, and do, thrive in disturbed habitats, but most of these – by the very nature of their ability to counter disturbance (by traits such as high dispersal and non-specific food requirements) – are not those of prime conservation interest. For species lacking these generalist features, many forms of increased landscape heterogeneity pose potential or actual threat. Small remnant habitat patches, isolated refuge areas for particular species, may need intensive management to maintain them. In Australia, one such example is the Eltham copper butterfly, *Paralucia pyrodiscus lucida*, near Melbourne, where all of the few populations occur on small isolated urban remnants (Canzano *et al.* 2007), with threats that overlap in generality but differ in detail across individual sites. For this insect and many others, site-specific management recommendations must be superimposed on more general considerations. In such small areas (of a few hectares or less), each needing continuing management, this butterfly is essentially conservation-dependent, with its survival depending wholly on the quality of these tiny inhabited sites. Its conservation is a long-term project, with considerable uncertainty over the eventual outcome.

A few studies, such as that on the marsh fritillary butterfly, *Euphydryas aurinia*, in central Europe by Anthes *et al.* (2003), have emphasised the need to integrate spatial variations in resource use and wider habitat quality with information on population structure. For this species, a suitable conservation strategy should incorporate conservation of a network of habitat patches and maximising local habitat quality on each patch by ensuring that the large host plants favoured for oviposition are maintained in situations of low vegetational diversity and no 'crowding' by grasses or other plants. Examples of such specific needs could be multiplied, but the pertinent generalities include:

1. that detailed knowledge of biology and requirements of both larval and adult stages of such insects constitutes the basis for informed conservation management;
2. that studies on a single population or site may not represent the full picture for any species, and provide only inadequate basis for extrapolation to other sites or populations;

3. that wider investigation of biology and requirements to demonstrate the ecological amplitude of the species across its regional or wider range may reveal trends not apparent from a single site;
4. that behavioural studies may augment understanding of how the species 'works' and help define the boundaries of the effective environment for the species; and
5. that knowledge of such variations in biology may further inform management. Insect behaviour may be unexpectedly complex, and aspects of dispersal (distance, landscape effects, difference between sexes and at different ages), resource-finding, mate-seeking, and others all integrate with (and are components of) patterns of habitat use in helping to understand the requirements of a species.

Implications of habitat fragmentation as a major cause of conservation concern for many insects are frequent, with the consequences centring on population isolation and increased vulnerability. However, although it is easy to suggest possible effects of this process for a given insect, it can be correspondingly difficult to validate and explore those effects in the field – not least because of inability to safely manipulate small populations or otherwise rare insects (p. 20). A background essay by Henle *et al.* (2004) elaborates many of the factors that may help to forecast susceptibility of species to habitat fragmentation. Very generally, they suggested that trends of population size (small size increases vulnerability), population fluctuations (high fluctuations associated with increased vulnerability), susceptibility to disturbance, microclimatic and other specialisations, patterns of resource use (largely linked to population size), low abundance within the habitat, and relative biogeographical position (related to area-wide effects such as large scale loss of grasslands) may all be important. These aspects should be considered in planning management at this wider scale.

Habitat management for insects (Kirby 2001) must encompass the duality of place and resources, and also the wider landscape context of these to counter possible deleterious effects of isolation and range reduction.

Summary

1. 'Habitats' for insects comprise places where a suite of critical resources occur in adequate supply, and environmental conditions are suitable for the species to persist. Very fine scale considerations may be involved

in defining optimal habitat for an ecologically specialised insect. Each such species may have very specific microclimate tolerances, food needs and, for some butterflies, also obligate mutualistic relationships with ants. Adults and early stages of the same species may have very different resource needs, so that conservation planning must heed the entire resource portfolio on which the species depends.

2. Such suitable habitats may be very scattered or widely dispersed as 'patches' within a landscape 'matrix' of inhospitable terrain. Such patches may appear isolated and, unless the insects can disperse adequately across non-suitable areas, are functionally so. Maintaining the capability of an insect to reach suitable habitat patches through landscape 'connectivity' is a management component of considerable importance in the conservation of many insect species.

3. Habitat models can be used to help define the suite of critical environmental features that characterise a good habitat for a species, and also to help predict the incidence of such habitats elsewhere. The parameters included in such models usually reflect correlation with species incidence or abundance, based on currently occupied habitats.

4. It is presumed commonly that insects on discrete habitat patches manifest closed populations, but studies on butterflies (in particular) have shown that many species manifest a metapopulation structure, whereby a series of patches within a landscape may be individually subject to rolling extinction–recolonisation cycles. In this case any local extinction may be 'natural' within the species' normal dynamics, rather than the cause for serious conservation concern.

5. Habitat quality and extent are a universal concern in insect conservation, and almost invariably a central focus of a conservation plan.

4 · *Current and future needs in planning habitat and resource supply*

Introduction: space and time in insect conservation management

Most current emphasis on insect species conservation is on attributes of 'space': the sites or habitats frequented by the species or on which they can be anticipated to occur within a very few years. Recovery plans normally commit support and action only for a limited period, and rarely extend beyond about a decade. However, conservation is also (just as importantly) about the longer-term future of the species as well as their present circumstances. Once any vital short-term measures to immediately safeguard the species have been assured, conservation planning may change to encompass the longer term. The dimension of 'time' is highly relevant, and satisfactory management must − as far as possible − anticipate a species' needs beyond the immediate context that might be presented. This may mean looking at a landscape context for sites and resources, or trying to forecast site and resource suitability beyond the immediate duration of a management plan. Focus on an individual site may be the immediate primary need for managing a sedentary or poorly dispersing insect species, but conditions that are entirely favourable at present may change dramatically in the future. Sustaining these can become increasingly difficult as those changes occur. Around 35% of the British butterfly species can persist in sites as small as 0.5–1 hectares (Thomas 1984). However, an individual habitat patch of such size may change quite rapidly through succession, and remain suitable for only a few years (or few insect generations) if left to its own devices.

Although such sites are often managed intensively to conserve insects, the increasing intensity of management over time may limit both the success and the amount of effort that can be continued. In such cases, persistence of the species depends on the availability of other suitable sites within the normal dispersal range of individuals and, as C. Thomas (1995) commented, insect species may be too sedentary to colonise such sites more than only a few hundred metres away. This high level of sedentariness is commonly not appreciated, particularly by non-entomologists, who sometimes assume that all winged insects disperse readily and over large distances. Studies on dispersal and other behavioural features may provide important clues to assessing effects of habitat loss through increased isolation of populations or subpopulations, and of the patches they frequent. This aspect of conservation has assumed greater urgency with the realisation that climate changes over the next few decades will influence many aspects of the tolerances and spatial coincidence of insects and their resources, with likely need to track these through changes in distribution across landscapes and to predict and provide for dispersal opportunities in the future. Many declines of insects have been attributed to loss of specific resources needed by immature stages. These resources are often more specialised, more spatially restricted, and apparently more limiting than those for the corresponding adult stage. Most immature insects normally disperse very little, particularly in comparison with the winged adult stage, and are thus 'tied' even more closely to their immediate surrounds and the local resources present.

Dispersal and connectivity

As noted above, even many adult insects, despite popular impressions to the contrary, do not disperse far, or may be prevented from doing so by features in the landscape. Even a few tens of metres of cleared ground may be a formidable barrier for some species. Many insects are unlikely to be able to negotiate such barriers. However, this is a very difficult topic on which to generalise, as a 'proven barrier' to one species may be ignored by another or even adopted as a conduit to facilitate movement. Simple 'dissection' of a habitat patch, for example by construction of a normal width roadway, may effectively divide a previously entire population into two discrete entities, but also allow other species to disperse and invade the areas along the cleared roadway. Such effects are indeed difficult to anticipate, but much relevant background to the effects of landscape features on insect dispersal has come from studies of butterflies in agricultural environments. A number of these were reviewed by New (2005a), and studies of movements of insect predators (such as ground beetles) and parasitoids (mainly wasps) into crops, as biological control agents for crop pest arthropods, have also contributed significantly to understanding of insect movement. A single example, from among the many that could be

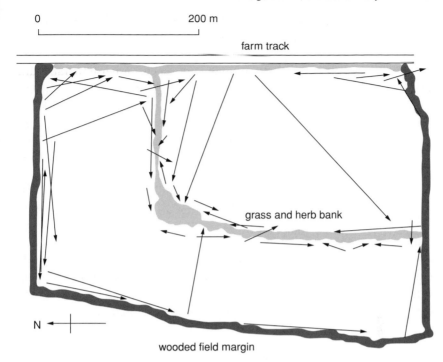

0   200 m

farm track

grass and herb bank

N

wooded field margin

Fig. 4.1. Movements of a lycaenid butterfly in an arable landscape: *Heodes virgaureae* in Norway. Each arrow represents the track of an individual butterfly observed over 15 minutes. Most butterflies associated clearly with edge habitats, and very few crossed borders, either the wooded margin or perimeter track, or the bank through the centre of the field (Fry *et al.* 1992).

cited, indicates some of the patterns that can be important to consider. Figure 4.1 shows the movements of a small lycaenid butterfly (*Heodes virgaureae*) in arable fields in southern Norway (Fry *et al.* 1992). Each arrow shows the path of a single individual tracked over a 15 minute period. The butterflies were associated closely with edge habitats and very few of them crossed borders such as the bank through the centre of the field or the field perimeter. Additional studies on this species showed that different structures of field margins had different 'permeabilities', features that differentially influence the proportion of butterflies passing across or through them. Height of vegetation was one such influential factor.

The major relevance of such studies here is to emphasise the importance of the landscape in insect species conservation, in facilitating or impeding connectivity between isolated populations or sites. Increasingly, the insects we seek to conserve are those that occupy small islands of suitable habitat in largely altered landscapes, and more or less isolated from other such islands. An important consideration for management, perhaps particularly for species that form metapopulations (p. 91), is to provide 'habitat networks', a number of suitable sites within the normal dispersal capacity of the species, so that each may at times support a demographic component of the population.

Conservation planning necessarily involves emphasising management of individual sites in many insect programmes, but the wider spatial perspective noted here is integral to sustaining many species.

Part of the practical problem over assessing the extent of natural dispersal an insect undertakes is indeed the size of the study area. As Franzen and Nilsson (2007) emphasised, this parameter of dispersal is extremely difficult to measure, because the rate and distance of dispersal is often underestimated in conservation studies, for two main reasons: (1) study areas are too small to allow insects their full dispersal potential; and (2) habitat patches used to estimate inter-patch movements are aggregated. In addition, for many insects of conservation interest, the numbers of individuals available for marking and study are necessarily small, and rarely exceed a few hundred at the most. Nevertheless, in the widespread absence of more sophisticated measures of dispersal, such as genetic markers, the more traditional study tools seem likely to persist for some years to come. In an attempt to estimate the size of the landscape required to make reasonable estimates of natural dispersal distances, Franzen and Nilsson (2007) studied two species of univoltine burnet moth (*Zygaena viciae, Z. lonicerae*) in Sweden, across a suite of 68 patches of flower-rich seminatural grasslands in a study area of 81 km². Substantial numbers of these diurnal moths were individually marked, with additional distinctive marks given on recapture, to trace their lifetime movements within this area, over two seasons. Altogether, more than 7000 individual moths were marked. The largest confirmed dispersal distance by an individual was 5600 m, and more than 100 inter-patch movements were detected. Franzen and Nilsson suggested that realistic estimation of the species' dispersal distances required a study area of at least 50 km². This has, in practice, rarely been available for most threatened species studied.

Much has been made of the values of linkages between habitat patches created by means of putative 'corridors' or more intermittent 'stepping stones' along which insects may move and avoid inclement intermediate areas. Some studies suggest that such linear features can be effective conduits even if they do not themselves comprise habitat for the species involved. However, and as Hess and Fischer (2001) emphasised, such structures can have a variety of functions. They may be conduits for passage without being suitable for more permanent residence; additional suitable habitat so that two previously separated habit patches or populations are, in essence, joined; filters, across which some species pass but others are unable to do so; barriers to passage; sources (reservoir habitats) for insects which can spread to other areas; and sinks which insects enter and are destroyed. For most species, we have little if any *a priori* knowledge of which role or roles may pertain. Nevertheless, any rational attempt to decrease isolation merits serious consideration as a component of wider management for conservation. Even if such putative linkages are not useful now, additional stresses posed by climate change or other threats in the near future may well change their role. Likewise, current resources for insects (and, of course, other organisms) need to be considered in relation to possible future changes in their availability and distribution. It thus becomes vital to attempt to predict what sort of changes may occur.

Occasionally, increasing dispersal opportunities in a landscape can be viewed as a disadvantage. Rather than promoting movement opportunity, it may be advisable to use knowledge of the insect's dispersal behaviour to create barriers in the landscape – for example, to prevent undue loss of individuals from small isolated populations when there is no other suitable habitat available within a reasonable colonisation distance. Knowledge of how the insect reacts to the edges of a habitat patch is particularly relevant. Several studies on butterflies have provided inferences similar to that from a study of *Maculinea alcon* in Belgium (Maes *et al.* 2004), in which many individuals turned back at the edge of their habitat, with this tendency increasing with height of trees there. Maes *et al.* consequently suggested planting tree rows around isolated sites. Their mark–release–recapture studies (see also Box 1.3, p. 34) showed that most butterflies dispersed normally for less than 500 m.

Future needs: climate change

It is entirely natural that conservation priorities should be driven by the most urgent concerns at the present time, but the future needs can not be excluded from any holistic planning for organisms such as insects. Many insects have distributions that appear to reflect rather narrow temperature or other environmental bounds, and the prospect of climate change is thus an important consideration for their future survival and distribution. The general principle of emphasising ecological gradients in future reserve design and designation is intuitively wise – so that, as examples, altitudinal and latitudinal gradients may to some extent counter the effects of climate change on otherwise isolated sites, from which the inhabitants are otherwise likely to be unable to move. At least in theory, corridors along gradients provide a means of tracking gradual changes in resources as their distributions change. However, planning for the moderate- to long-term future of those insects that we perceive to be already occupying only the most extreme environments, or very narrow climate regimes, and which we see as having 'nowhere to go' is clearly difficult. Locally endemic high altitude dwellers are one such context of concern. For example, the alpine satyrine butterfly *Oreixenica latialis theddora* is endemic to Victoria's Mt. Buffalo plateau, and occurs on native grasslands near the highest parts of this isolated elevated area. If its distribution is indeed limited by need for cool temperatures, which remains to be proven experimentally but is assumed widely to be the case, it may well be doomed as the local alpine environment warms.

Hanski and Pöyry (2007) emphasised that species may be able to move their ranges only if there is enough habitat in the landscape, and this is not too fragmented. For British butterflies, Warren *et al.* (2001) noted that generalist species (with a large amount of suitable habitat) showed range shifts with climate change, but specialist species that depend on scarce and highly fragmented habitat patches did not do so. If this proves to be a more general pattern, the more ecologically specialised species (including most insects of current conservation concern) may suffer more than co-occurring generalist species in response to climate change.

The possible consequences of future climate change are thus now a fundamental consideration in planning conservation of any insect or other species, with three broad categories of responses long recognised.

1. Extinction, if climate changes to eliminate the specific regimes needed by a species throughout its current or potential range.
2. Adaptation within the current range, if the species is sufficiently 'flexible' to thrive under the changed conditions.
3. Migration, so that the species' range changes to encompass areas now expressing the tolerable climate regime, and manifest in range expansion and (sometimes), loss from part or all of a former range.

In their early perceptive anticipation of the effects of climate change on the world's biota, Peters and Darling (1985) identified a suite of species cohorts that might be rendered especially vulnerable. These are

(1) species that have narrow or geographically localised distributions; (2) habitat specialists; (3) alpine species; (4) species that disperse poorly; (5) species that are genetically impoverished; and (6) species with peripheral and/or disjunct distributions. Many insects of specified conservation concern, and myriad others, fulfill one or more of these conditions, particularly if they are also in some way 'rare' (p. 28). The numerous narrow-range endemics, for example, must be considered vulnerable to climate change unless it can be proved otherwise. However, one almost universal unknown factor is the effect of rate of change, rather than just its extent. A few insect examples give us some clues, as representatives of well-mapped faunas whose distributional changes have been assessed reliably over many decades. For such taxa, changes in both geographical range and more local aspects of site/patch use may be evident within relatively short periods.

Most detailed information is available on the northward expansion of many butterflies and dragonflies in Britain during the last few decades, and the relationships of this to climate change. There, concerns have been expressed for more northern species, many colonies of which have been lost. Hickling *et al.* (2005) examined distribution trends in all 37 non-migratory Odonata occurring in Britain, using records for 1960–1970 and 1985–1995. Of these, 24 species reach their northernmost limits in Britain, four have their southern range limits there, and nine occur throughout Britain. All but two species increased range size over the period, and all but three shifted northward. Northern species tended to be displaced northward from their southern range margins. Although those shifts were correlated with increased temperatures, there is also possible effect from improved water quality over much of the area during the same period, so that range expansion could also be due to improved habitat quality. There is also a possible effect of increased sampling efficiency, because the number of recorders increased more in northern Britain than in the south. Thus, even for such relatively well-documented cases, possible alternative or complementary interpretations of change are possible. However, the consistency of trends indeed implies that the main factor involved is temperature change. In addition, for many British dragonflies, phenological changes have occurred over the 45 year period from 1960 to 2004 (Hassall *et al.* 2007) so that flight seasons now differ somewhat from those of earlier decades.

In short, distributional effects attributed with reasonable confidence to climate change are already evident, and are not simply something for the future. Local changes in resource use also occur. Formerly the silver-spotted skipper butterfly, *Hesperia comma*, in Britain, depended wholly on the hottest south-facing habitat patches with very short turf; it now breeds also in somewhat longer turf and other aspects of its habitats. The extent of thermally suitable habitat has indeed increased at the cooler northern edge of its distribution, leading to increased fecundity, larger populations and reduced susceptibility to extinction of populations. Density within habitat patches has also potentially increased – either from increased patch quality or increased effective area. Broadening of the skipper's realised niche results in increased relative size of habitat patches, and increased connectivity between them (Davies *et al.* 2006). *H. comma* shows a metapopulation structure, within which most individuals usually disperse over only about 50–100 m, although one marked individual moved slightly more than 1 km (Hill *et al.* 1996). This Holarctic species is widely distributed and, in many places, common: it is taxonomically complex in North America (Forister *et al.* 2004).

A second butterfly example of changed resource quality involves the Richmond birdwing (p. 61) in Australia. Prolonged drought over the past few years has been associated with increased toughness of the foliage of the caterpillar food plant vine, together with increased concentrations of foliar toxins, which together render the plants inedible so that, even in the apparent presence of plenty of food, caterpillars starve to death (D. Sands, pers. comm. 2007). In some places in southern Queensland, individual people have been watering and fertilising planted vines so that they produce soft, edible foliage. This is a simple and effective conservation measure but clearly, in a climate of likely decreasing rainfall, may not counter this effect in natural communities. There, persistent drought may effectively remove the major food supply for caterpillars, and counter the major conservation measure used at present of continuing to plant vines throughout the butterfly's range.

Several modelling and sampling studies for insects in Australia contribute to understanding climate effects on possible changes in distribution. Two rather different approaches are complementary in enhancing our knowledge.

1. Sampling the same insect species or group of species in a widespread habitat, such as on a widely distributed putative host plant along a latitudinal or altitudinal gradient. Andrew and Hughes (2005) enumerated phytophagous bugs (Hemiptera) at four stations along the 1150 km eastern coastal distribution of a wattle, *Acacia falcata*. Their samples collectively comprised 98 species of bug. Overall species

richness was less at the southernmost latitude, and only 10 species were collected at all four sites. Many (57) of the species were categorised as 'rare', as they were found at only one site, at one time, and represented by singletons. This circumstance is very common in multispecies samples of insects, and interpretation is hindered by the fact that many of these scarce species, inferred to be ecological specialists, may be those of greatest conservation interest but also those most difficult to evaluate in any reliable or semi-quantitative way. Excluding those species of Hemiptera from their analyses, Andrew and Hughes (2005) recognised four groupings, based on distribution and host plant spectrum: cosmopolitan (more than one *Acacia* sp. and more than one latitude; generalist (found on more than one *Acacia* sp., but at only one latitude); specialist feeders (only on *A. falcata* at one latitude); and climate generalists (only on *A. falcata* but at more than one latitude). Andrew and Hughes hypothesised that understanding current distributions of insects in this way may facilitate prediction of future resources for such phytophagous insects. However, several caveats were noted for this: (1) that increasing temperature will indeed be the most important factor influencing species distributions; (2) that all species were sampled to characterise the assemblage; and (3) that an insect collected on the host plant was also feeding on it rather than being simply a tourist. Suggestions include that the specialist species may be the most vulnerable to local extinction, as changes in distribution of the insect and its host plant must coincide. In contrast, cosmopolitan or generalist species retain more ecological options for changed conditions. Very similar conclusions resulted from a parallel survey of beetles (96 phytophagous morphospecies) along the same gradient (Andrew & Hughes 2004).

2. The second approach involves modelling, applying information gained on the climatic envelope or environment currently occupied by a species and using this to forecast where those conditions might be available in future. Limited field data are sometimes available to document the reality of this approach, with the prevailing beliefs that shifts in distribution toward higher latitudes and altitudes may well represent responses to warmer temperatures, with the reverse trend much more unusual (Parmesan *et al.* 1999). For the well-studied New World butterfly *Euphydryas editha* (Edith's checkerspot), population extinctions along the west coast show a cline with altitude and latitude, and populations in Mexico are about four times more likely to have become extinct than those in Canada (Parmesan 1996).

The initial 'climate envelope' assessed for any species may include little more than information on the maximum and minimum temperatures it experiences, with the assumption that mean annual temperature adequately approximates 'climate'. This approach, however simplistic, is practicable as an initial guide, with the additional assumption that climate is a determinant of the species' distributional range, and that the species achieves its limits of normal tolerance over that range. For many plants, additional climatic features such as precipitation and soil moisture may also be important determinants of range – so that these features become vital also for associated insects for which those plants are a critical resource because both their presence and their quality may be affected. Such simplified climate envelope estimates allow some initial appraisal of species' vulnerability to change, but adding a wider range of relevant factors adds substantially to understanding.

Beaumont and Hughes (2002) used the model 'BIOCLIM v.5.0' (involving 35 climate parameters) to evaluate the climatic envelope of 77 Australian butterfly species with limited (less than 20°) latitude range. Twenty-four of these species were investigated further to model potential changes in distribution with four possible climate change regimes projected to 2050. Life history information was integrated with those models to identify which species might be especially vulnerable to climate change.

Seven species were indeed identified as particularly vulnerable. Five had very narrow climatic envelopes (with mean annual temperature range spanning less than 4 °C), with scenarios revealing that mean seasonal temperatures by 2050 may exceed the values to which the species are exposed at present, and the other two species lost large proportions of their distribution range under all climate change scenarios projected. These species included specialists, mutualists and poor dispersers, and all have narrow current distributions. They were considered unlikely to be able to change distributions to track either a changing climate or changing distribution of the host plant. Four of the seven are myrmecophilous Lycaenidae, with the additional complication that they must also track their mutualistic ant species.

Despite the generalities indicated above, species react to particular environmental changes in individualistic ways and, as Crozier (2004) has emphasised, 'we are far from a detailed understanding of range shifts'. For Edith's checkerspot, noted above, climatically extreme events have directly caused local extinctions. Such events as drought, heavy rains, snowfall and extreme temperatures are all implicated in this (see Parmesan *et al.* 2000; McLaughlin *et al.* 2002), but tend to occur randomly. The

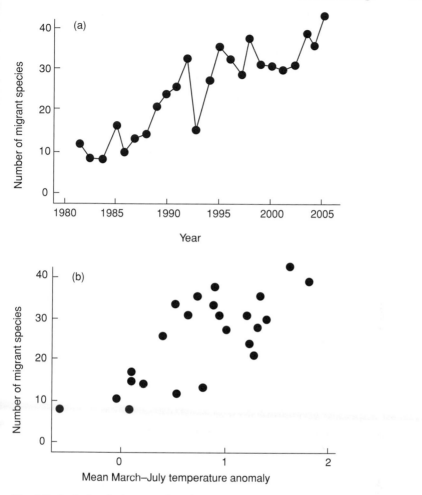

Fig. 4.2. Arrivals of migratory Lepidoptera in southern England, as a possible consequence of climate change. (a) The number of migrant species recorded each year at Portland Bird Observatory; (b) the relation between the number of migrant species recorded each year and the mean March–July temperature anomalies in southwestern Europe (see text, after Sparks *et al.* 2007).

trend with long-term climatic change is to create bias toward colonisation or extinction in different parts of a species' range so that large-scale distribution shifts occur over time.

Monitoring the number of species of migratory Lepidoptera at Portland Bird Observatory in southern England over more than 20 years to 2005 revealed increases over this period (Sparks *et al.* 2007; Fig. 4.2a). The number of species is linked strongly to rising temperatures in

Table 4.1 *Species borders: associations between three population characteristics and possible factors responsible for species borders*

Factor affecting border	r_m	density	condition
Climate	ramp	ramp	ramp
Substrate	step	step	step
Resource not modified by organism	ramp	ramp	ramp
Resource used by organism	step	ramp	step
Facultative predator, parasite, or pathogen	ramp	ramp	ramp
Obligate predator, parasite, or pathogen	(difficult to predict: very varied interactions)		

Source: Caughley *et al.* (1988); see text for explanation.

southwestern Europe, with suggestion (Fig. 4.2b) that a 1° rise in temperature was associated with an additional 14.4+/−2.4 migrant species. Most of the 75 species recorded over the period have had to negotiate a minimum of 150 km of open sea to reach Portland. This example is thus particularly interesting in indicating a phenomenon rather different from simple range extension across land, without any such intervening obstacle.

Climate prediction models, exemplified above, seek to determine and forecast a species' geographical range by defining its 'climatic envelope' and predicting where this will occur in future. Typically, such analyses reveal a 'core region' where conditions appear entirely suitable, and declines toward peripheries of a range as conditions change. As Hoffmann and Blows (1994) noted, marginal populations sometimes appear to survive under very adverse conditions, and the borders of a species' range can be influenced by many different environmental factors (Table 4.1). They cited the study by Caughley *et al.* (1988) on kangaroos, in which a method was proposed to compare central and marginal populations in order to identify the relative importance of biotic and abiotic factors, and to distinguish between the effects of climate, resource use and presence of natural enemies (predators, parasites) in determining distributional limits. Three parameters employed were density, condition (some measure of fitness), and rate of population increase (r_m). However, for most insects the fundamental parameters of intrinsic rate of population increase and the 'condition' of individuals (estimated by indices such as fecundity and fertility) are not available, and are unlikely to become so easily, so that density (broadly, abundance) becomes the main measurable feature. The two main trends between the centre and periphery of a range are 'ramp' differences (that is, progressive changes) and 'step' differences (sudden changes, such as presence/absence). Hoffmann and Blows noted a variety of ecological and evolutionary hypotheses influencing species distributions. In practical conservation, our concerns tend to emphasise the former, but understanding is augmented considerably by incorporating evolutionary considerations as well. Not least, some understanding of the genetics of stressed populations, and the variation within populations, can guide conservation practices such as translocations and re-introductions (p. 173) effectively.

Box 4.1 · *Bioclimatic variables and modelling insect distributions*

'Bioclimatic modelling' involves correlations of a variety of climatic and distributional factors (based on features of sites from which the focal insect species has been recorded) to characterise the species' suitable climatic regime and to predict where this might be found elsewhere, or in the future, in conjunction with climate change models. It thus defines and predicts the climatic limits of the species, and is an important tool in interpreting biogeographical patterns. Several such models were developed to help predict the distribution of alien pest insects in new countries, as well as whether potential classical biocontrol agents might operate effectively in controlling them.

For the endemic alpine Ptunarra brown butterfly (*Oreixenica ptunarra*) in Tasmania, McQuillan and Ek (1997) used the program 'BIOCLIM' to summarise characteristics of all inhabited sites, to demonstrate the occurrence of a longitudinal cline inferred to be related to the increased efficiency of thermoregulation for adults in more marginal habitats. The species occurs in habitats that differ considerably in annual rainfall, but far less in annual temperature range. BIOCLIM is based on a range of climatic parameters derived from numerous sites to produce 'climate surfaces' based on 35 variables related to temperature and precipitation indices, and is applicable when only presence/absence data at sites of known latitude/longitude and elevation is available. Another popular program, CLIMEX, describes taxon responses to temperature and moisture, using either or both of distributional and biological data to derive a 'population growth index' and 'stress indices' (cold, hot, wet, dry). These are combined to give an 'ecoclimatic index' of overall suitability of any area for permanent occupation, and to rank sites for this compatibility.

As another example, Steinbauer *et al.* (2002) examined institutional collections of coreid bugs and obtained locality data from their labels. The importance of this approach in providing locality data for rare species is that considerable distributional data may be accrued without the need for intensive field surveys. These data, of course, may still not be comprehensive, but may considerably augment the more immediate information available and the outcome help to focus survey efforts for additional populations. Combination of museum specimen data and climate data may thus augment understanding of patterns that result from newly gathered field data alone. However, one caveat of museum records is that the specimens available are often unlikely to represent a species' entire range, because they are more likely to have been collected sporadically rather than as a result of systematic surveys (Beaumont *et al.* 2005).

Experimental 'transplants' of insects to test their immediate capability to survive beyond their current natural range may not allow for the possibly more gradual adaptations that could occur in nature. Crozier (2004) sought to explore whether temperature might influence distribution of the North American hesperiid *Atalopedes campestris* by translocating individuals at sites along a gradient of 3 °C beyond the current range. Caterpillars of this generalist skipper butterfly feed on many grass species, and Crozier suggested that ample natural habitat had long occurred outside its range, such as in rural and suburban areas, and should perhaps have been colonised already if the environment was suitable. Eggs were transplanted and the release points later caged to confine emerging butterflies. Developmental time was slower outside the normal range than within it, but survivorship, fecundity and predation pressure did not differ significantly. The current range limits may be defined by a combination of summer and winter temperatures, with slower development from cooler temperatures also reducing the number of generations possible (Crozier 2004).

Range shifts (either geographical as above or altitudinal: see Konvicka *et al.* 2003, for European butterflies) are only one of the possible consequences of climate change. Perhaps as important to consider are likely phenological changes, whereby conventional patterns of species appearance may change. Warmer conditions may be reflected in earlier spring/summer appearance of species, which may then depend on parallel trends in their critical resources. Although noted so far mainly for northern hemisphere butterflies (examples: Roy & Sparks 2000, Britain; Forister & Shapiro 2003, California), this trend is likely to be far more widespread, and to occur in many groups of insects. Broader background to the various correlational studies involving insects and climate change, and a number of studies on individual species, are summarised by Wilson *et al.* (2007).

Box 4.2 · *Climate change and narrow-range endemic insects*

The opportunities to incorporate planning for climate change into conservation of narrowly distributed insects will clearly vary with the particular nature of that distribution for each focal species. Extending the principles of distribution noted earlier (Fig. 1.3, p. 23), any one of about five categories or patterns may be relevant (Fig. 4.2.1) when the known distributions of species that command attention are superimposed on the possible range of latitude or altitude in the region. Thus, a species may already be known only at an extreme of such a gradient (Fig. 4.2.1, number 1) (such as some alpine taxa: see New and Sands (2002), on Australian butterflies). Perhaps more commonly, a species is known from a single population or site in a presumed intermediate part of the possible range (Fig. 4.2.1, number 2), so may have capability to 'move'. Third, a species may be found with populations or sites separated but loosely grouped within a broad range (perhaps by up to several hundred kilometres of latitude, or several hundred metres of altitude). In contrast, populations or sites may be grouped much more tightly (Fig. 4.2.1, number 4) to constitute a distinct local endemic with a strongly concentrated and well-defined range. Last (Fig. 4.2.1, number 5), a species may be known only from isolated populations at or near the extremes of its possible range. This scenario is perhaps the most difficult to interpret, and may commonly be assumed to be associated with either (1) extensive loss of intermediate populations to leave these remote remnants or (2) need for taxonomic clarification, so that the two entities may in fact be different taxa and each parallel

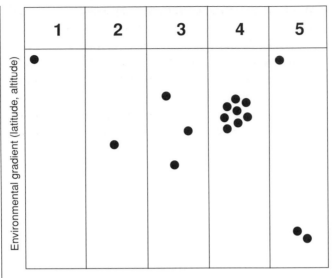

Fig. 4.2.1. Distributional patterns of narrow-range endemic insects within a possible environmental gradient, such as latitude or altitude (see text).

the first category noted above. In any case, their high separation may accord them status as 'significant populations'.

It is commonly presumed that category 1 taxa have 'nowhere to go' because they already occupy extreme environments that may be eliminated as conditions change, so that conservation may be futile. Alpine species may already be on the highest land available, for example. Commonly, conservation is directed at the site(s) in the hope that the species may be capable of adapting to a changed climate regime when its resource needs are assured. The assumption depends on the taxon being restricted due to intolerance of less extreme or different regimes (for example, of temperature) and, in almost every case, this remains unproven.

Wider options may be open for category 2 and category 3 species. Unless these are extremely specialised in relation to an optimal climate regime, a category 2 species may have capability to expand along the gradient in either direction from the central distribution, so that acquiring, safeguarding and rehabilitating/restoring additional sites within a relatively broad range, for either translocation or natural colonisation, may be a viable conservation action. A similar approach may be adopted for category 3 species, perhaps with potential to expand beyond one or both current range extremes. Category 4 species may prove to be climatically limited, as reflected in a 'tight' concentration, but the options for category 2 species merit consideration for these. Any such suggestions presuppose that the reasons for the current distribution of the species are understood: in most cases that understanding is based on presence or quality of biological resources, rather than of optimal climate regimes.

Modelling studies that imply future range changes through poleward movements are often based on the presumption that climate change will both present opportunity to expand from the current range, and render parts of the current range unsuitable. These inferences are perhaps best interpreted in the context of a comment by Beaumont and Hughes (2002, p. 969): 'Bioclimatic models do not represent forecasts of future distribution, but rather provide an indication of the potential magnitude of the impact climate change may have on these species distributions'.

The wider implications of effects of future climate change on distribution patterns and survival of many insects can scarcely be conjectured but, wherever possible, these should be considered in planning management. For example, the synchrony (in both space and time) between insects and their food plants may become disrupted, so that insects no longer have easy access to critical resources, and in turn many plants may lose access to specific pollinators. Enforced adaptations to changed seasonal development may affect diapause regimes so that current 'temporal refuges' are no longer available or effective. To date most of our inference is on gradual changes in distribution wrought by changing temperatures, but wider changes in weather patterns (such as increasing drought periods) may impose more abrupt changes in insect survival and reproductive capability. Critical resources or critical sites may simply disappear, become unsuitable, or otherwise diminish in accessibility. Wilson et al. (2007) noted that the overwintering sites for adult *Danaus plexippus* in Mexico (p. 29) may change severely in character (from their current cool dry conditions to cold and wet conditions) within 30 years, for example. One wider inference from Wilson et al. (2007) is that climate change is likely to increase the vulnerability of most species that are already threatened. With such a premise, optimal management can not be planned at present, except in heeding several rather basic considerations. These include increasing connectivity within landscapes to facilitate movement by relatively sedentary insects and so as otherwise not to impede dispersal, emphasising conservation measures along ecological gradients, and maintaining habitat heterogeneity to facilitate choices for local adaptations such as microclimate and host plant selections.

Summary

1. Much habitat management for insects naturally involves characteristics of 'place', with the suitability, security and carrying capacity of suitable places for a species a major focus.
2. Landscape features are of critical importance in promoting or impeding dispersal of insects between suitable patches. Contrary to common belief, many ecologically specialised insects naturally move only over short distances, so that 'barriers' to dispersal can easily and unwittingly be created by human activities, and increase isolation of populated patches.
3. However, in addition to current spatial considerations, climate change predictions imply that likely changes in range for many insects must be anticipated, with much of their current range becoming unsuitable for continued occupation. Distributional shifts include poleward and higher altitude movements, with synchrony between insects and their resources (such as food plants), and timing of seasonal appearance also affected. Many such changes are extremely difficult to predict and, even with the best climate prediction models that exist, considerable uncertainty persists.
4. The options for conserving highly specialised insect species, many of which are not good dispersers (so may not track changing resources in their environment) or adaptive to changed temperature or precipitation regimes, may be bleak. Possible counters to local extinctions by preparation of additional habitat patches elsewhere within the predicted ranges need to be considered carefully.

5 · Beyond habitat: other threats to insects, and their management

Introduction: key threats to insects

Habitat change is the predominant threat to many insects and is often that most easily observed. It is potentially universal. Countering and repairing habitat changes that have occurred through loss or degradation is therefore the primary focus of conservation management for almost all species of concern. However, a variety of other concerns may arise, and in some cases one or more of these become central themes for management to address. Each has had major importance in some insect management plans, and they are noted here to exemplify some of the problems that may arise, and to ensure that they are not overlooked in planning. However, unlike habitat considerations, they may not be a factor in every programme. Conversely, they may be highly important. When they are present, as with changes to habitats, they may either cause direct mortality or affect the quality of the environment within which the species dwells. Some causes of direct mortality of insects are very difficult to anticipate and may not be obvious until closely investigated. USFWS (2001) summarised information for a dragonfly to suggest that mortality from direct impacts with vehicles or trains may contribute to reducing population size, for example. In Illinois, railways pass close to or through three of the sites supporting the largest populations of Hine's emerald dragonfly. At two of these sites, train speeds are reduced to only *c.* 6–10 km h^{-1} during the dragonfly flight season (Soluk *et al.* 1998), as a measure to reduce impact mortality. As an even more unusual threat circumstance, even the name of a species may lead to problems; the so called 'Hitler beetle' (*Anophthalmus hitleri*) is reportedly threatened by collectors pandering to the lucrative Nazi memorabilia market, and poaching specimens from their restricted cave habitats in Slovenia (Elkins 2006).

Alien species

Species that may be regarded broadly as 'unnatural' occur in many communities, and many of them are 'exotic' in the sense that they have either been transported there by human agency, sometimes deliberately, or have arrived by accident, sometimes from distant parts of the world. Should these species invade natural environments, many and varied interactions between alien animals and plants with native insects can occur. They are often perceived as a threat. In other cases they may constitute an additional resource, but conflicts of interest or priority can then arise and lead to debate over needs to suppress the invading species.

Some of the more distinctive and widespread recent concerns for the wellbeing of some native insects have arisen from the effects of non-native plant species that are adopted as oviposition sites or larval food plants, in a number of different contexts. Consider three of these:

1. The introduced South American vine *Aristolochia elegans* (Dutchman's pipe) has been planted extensively as an ornamental plant in gardens in southern Queensland and northern New South Wales. It continues to spread into natural bushland in Australia. *A. elegans* has become a weed in many habitats occupied previously by the native *Pararistolochia praevenosa*, the major natural food plant of caterpillars of the Richmond birdwing butterfly, *Ornithoptera richmondia*. *A. elegans* is attractive to female butterflies for oviposition, but its foliage is toxic to the eclosing caterpillars. The spread of *A. elegans* is therefore a significant threat to *O. richmondia*. The butterfly's management plan includes a continuing programme of removal of the vine, accompanied by enhanced planting of *P. praevenosa* (Sands *et al.* 1997; Sands & Scott 2002). In this case, reduction and removal of the exotic decoy plant is a critical component of the butterfly's conservation throughout its range.

2. Chilean needle grass, *Nassella neesiana,* is an invasive weed of native grasslands in southeastern Australia, and displaces native grasses such as *Austrodanthonia* spp. in such areas. As a 'Weed of National Significance' in Australia, it is a target for control and eradication in such areas. However, recent observations (Braby & Dunford 2007) imply that it might be adopted as a food plant by caterpillars of *Synemon plana* (p. 30), in which case it might be an important resource in conservation of the sun-moth in areas where native food plant grasses are sparse. The dilemma of encouraging noxious weeds in conservation of a threatened native insect poses interesting questions of priority,

public relations and permissivity. In this case, the weed appears also to be displacing native grasses of independent conservation value.

3. Weed invasions of prairies in the central United States have been stated to threaten the Dakota skipper, *Hesperia dacotae*, throughout its entire current range (USFWS 2005). Species such as Kentucky bluegrass (*Poa pratensis*), smooth brome (*Bromus inermis*) and Canada thistle (*Cirsium arvense*) rapidly become dominant and replace both larval food plants and adult nectar plants. The reasons for concern relate in part to seasonality: *Poa* and *Bromus* become senescent in late summer, at the time when skipper caterpillars need palatable fresh grasses to eat.

Effects of invasive plants sometimes extend well beyond the insect's feeding environment and replacement of natural food plants, as above. Severns (2008) noted that they may also change the structure of an insect's landscape and render it less likely that normal behaviour patterns will persist. He noted that changes in vegetation structure may degrade the quality of butterfly basking sites, and so also interfere with thermoregulatory behaviour and perhaps in turn affect a wide range of other behaviour necessary for normal existence. Likewise, overgrowing of natural, low-growing food plants may render them difficult or impossible to find, so that they are not then apparent as potential oviposition or feeding sites. The rather subtle case of food plant quality change for *Candalides noelkeri* (p. 95) may not be particularly unusual, although such cases of shading effects have not been widely studied. For Fender's blue (*Icaricia icarioides fenderi*), Severns (2008) found that a predominant invasive grass (*Arrhenatherum elatius*) in its prairie habitats in Oregon is 2–3 times as tall as other grasses. Clipping of *A. elatius* to the height of native grasses led to considerable increase in oviposition, more frequent basking by butterflies in cleared areas, and changes in butterfly dispersal behaviour. Such invasions may be 'an insidious form of habitat degradation for grassland Lepidoptera worldwide', but have gone largely unremarked (Severns 2008). Any reduction in host plant apparency and accessibility from alien plant increases or competition should be considered likely to be detrimental to the species of conservation interest, and attract management attention. Many other cases could be cited. Eradication of undesirable alien weed or insect species must be undertaken with consideration of any harm likely to be caused to the focal insect or its needs – which may include food plants closely related to the weed. Some relatively economical general weed control methods, such as use of a broad spectrum herbicide, may therefore not be available and more selective methods such as individually poisoning plants or physical weeding may be needed. Any such exercise can be laborious (i.e. expensive) and may also need to be repeated several times or over several seasons to ensure an adequate level of suppression, even if total eradication may not be possible.

Considerable debate has ensued over the non-target effects of insect predators and parasitoids introduced as classical biological control agents for agricultural and forestry pests. Particularly on isolated islands, these have been implicated strongly as threats to native species of insects, and the varying points of view continue to be advanced over their roles and the relative priorities that arise (summaries in Lockwood *et al.* 2001; New 2005a, for examples). Some such exotic agents have dispersed into natural vegetation systems, so are invasive and sometimes occur far from the cropping areas in which they were first released. However, perhaps particularly for insects in remnant habitats within agricultural arenas, such organisms should be sought as part of conservation monitoring and threat definition, and their possible interaction(s) with the target species clarified wherever possible, both by review of all information available and by practical investigation. It is likely that for at least some of these agents published information on host or prey range will be far greater than for any equivalent native species in the same area. If plans emerge to introduce any such 'natural enemies' into the vicinity of a known threatened insect species, more detailed screening or survey may be needed. However, it is well known that laboratory screening tests to determine or predict parasitoid or predator host or prey range can be misleading in relation to field choices. The predatory ladybird *Coccinella septempunctata* would eat eggs of the endangered lycaenid *Erynnis comyntas* in North America when its normal aphid food was not present, but this behaviour has not been observed in nature (Horn 1991). When 'stressed' or deprived of choice, many insects will eat or attack species that would not be taken normally in nature.

Invasive alien (usually exotic) social Hymenoptera (mainly ants and vespid wasps) are viewed widely as among the most serious candidates as potential threats to native insects, and wherever possible must be controlled in a threat prevention or abatement programme. Eradication of invasive social Hymenoptera is difficult, but likely to become a conjoint exercise because of their widespread threats to other animals and ecological interactions, so that any specific threat to an insect will be only a part (perhaps a small part) of a threat portfolio they pose. In some cases

Box 5.1 · *Effects of invasive tramp ants on native ant species*

Several species of ant have been distributed widely around the world and are known as 'tramp ants', with potential to disperse into natural environments within their climatic tolerance range and to interfere with and displace native invertebrates. They also disrupt a variety of ecological processes and are viewed collectively as a severe threat. Six such alien species are designated in Australia as of national threat significance and a 'threat abatement plan' for these has been prepared (Commonwealth of Australia 2006). Two species are pervasive in southern Australia, as below, and others are more predominantly tropical.

The Argentine ant (*Linepithima humile*) and the big-headed ant (or coastal brown ant, *Pheidole megacephala*) have both been implicated as a threat to native ant species in Australia, as aggressive colonisers. One of the genera apparently disrupted by *L. humile* is *Notoncus* (Walters 2006), whose conservation interest extends to its being the mutualist associate of caterpillars of the Eltham copper butterfly (p. 206), so that its loss may put the butterfly at additional risk.

Reasons for such displacement are not wholly clear, but may involve aspects of predation, competition and influences on the associations between native ants and honeydew-producing Homoptera. May and Heterick (2000) suggested that *P. megacephala* 'simply overwhelms other ant species', in a study that demonstrated (in Perth, Western Australia) that gardens with *P. megacephala* had fewer native ant species than gardens from which it was absent.

Similar inferences of the effects of tramp ants in many parts of the world (Holway *et al.* 2002) imply that, should any of them be found on sites where other insects are being managed for conservation, the ants should be considered a likely threat and steps taken to suppress or eliminate them.

conflicts of interest may arise, however. Introduced pollinators such as bumblebees, and extensive feral populations of honeybees, are viewed commonly as threats to native pollinating insects and have been implicated in their declines. Such effects are often very difficult to quantify or confirm but, equally, are hard to deny. The European bumblebee *Bombus terrestris* (naturalised in New Zealand, where it was introduced as a pollinator, for more than a century) was first discovered in Australia in Tasmania in 1992 and has now spread throughout much of that State.

Among other attributes, the bumblebee forages at lower temperatures than native bees, and is thus able to reduce the supply of nectar available to them (Hingston 2007). In the neighbouring mainland state of Victoria, introduction of B. *terrestris* is listed formally as a potential threatening process for several reasons, including (1) its potential to pollinate a suite of exotic plants and thereby facilitate the spread of some exotic weeds; (2) possible competition with native nectar-feeding animals; and (3) leading to possible decline in seed production of some native plant species. Formal proposals to introduce this bee to the Australian mainland are for confined (glasshouse) pollination services, but escape of queens to the outside cannot be prevented fully. Conflicts will continue to arise over differing priorities and perceived differences in the roles of an array of alien species.

The cases cited above involve deliberate introductions. Many species can be introduced inadvertently, in ways that may not be intuitively obvious. The effects of such species in the receiving environment may not be at all clear, even if the species involved are recognised. Large numbers of living scarabaeoid beetles are imported to Japan from parts of southeast Asia, for pets. The beetles are accompanied frequently by living canestrinid mites; Okabe and Goka (2008) expressed concerns that those mites could become established in Japan and transfer to native beetle species, to which they could represent a threat. The biology of the mites, which live in the subelytral cavities of the beetles, is almost wholly unknown. Some are thought to feed on host exudates, but others may be truly parasitic.

Larger animals, such as rodents, have also played significant roles as threats to native insects, most notably on isolated islands, and in New Zealand where spectacular insects such as weta (p. 63) have evolved in the absence of mammalian predators. Near-loss of the Lord Howe Island stick insect (p. 182) is also attributed to rat predation. Eradication of rats and mice, and occasionally of other vertebrates, is a critical component of site security for many insects, perhaps most notably in preparing sites for re-introductions or translocations. Stocking of water bodies with exotic fish for recreational angling might occasionally cause harm to Odonata and other aquatic insects (Suhling 1999), and care may be needed to control this on water bodies subject to management for such species. The presence of the mosquito fish (*Gambusia*), introduced widely as a biocontrol agent, has been correlated with decreased richness of Odonata in Australia (Davis *et al.* 1987) and Hawai'i (Englund 1999).

Part of a routine protocol for insect recovery could usefully be to include considerations of effects of all invasive species likely to be encountered, particularly those species documented as harmful elsewhere, and to seek advice on those effects. The main facets of this should be to:

1. Determine the presence of any such species on sites of conservation importance for the insect.
2. Determine the presence of any such species within the general area and from which they might easily spread to the conservation sites.
3. Determine what preventative measures, such as ongoing inspections or quarantine, may be needed to detect if and when such species arrive from 2.
4. If present, explore need and methods to suppress or eradicate them. If likely to invade, prepare a contingency plan for this.

Strategies for eradication of invasive species are many and varied. For animals they may involve trapping (live or dead), poisoning (with due attention to specificity and avoidance of non-target effects, below), shooting (for larger species such as foxes and cats, but often inefficient after initial reduction of populations), or disturbances such as raking warrens or destruction of nests or burrows. For plants, options may include hand pulling or weeding (laborious, but an activity in which volunteer groups may be able to participate), poisoning (with care as above, and sometimes involving topical application following slashing), or cultural methods such as mowing. For many of the more important species involved, suppression or eradication methods are likely to have been established during previous exercises, but others may need a more innovative approach, not least to ensure that the methods are integrated with the biology of the species we wish to conserve.

Pesticides

J. Thomas (1995b) raised the intriguing point that there had by that time been rather few studies of effects of pesticides on non-target insect species, because of the much greater threat they posed to bird populations, through which concerns some aspects of the major problems of pesticides in the open landscape were addressed.

Broad spectrum pesticides should not be deployed on sites where a focal species occurs, except under well-considered and responsible circumstances. Remnant habitat patches in agricultural ecosystems may be particularly vulnerable to spray drift from aerial or land-based pest management operations, and run-off to nearby water bodies may also pose a threat. Not all these may be obvious, and any chemical contaminants reaching water bodies may cause concern in particular cases. In addition to recognised pesticides, others such as orchard, agricultural or horticultural chemical run-off, or leachates from landfill or mining activities upstream, may need to be considered. For a few insect pest species, area-wide management may include spraying of substantial areas of ground, perhaps including reserves, and effective communication is needed to ensure that areas valuable for particular insect species are not inadvertently affected. Likewise, pesticides to control aquatic pests may need to be considered carefully: Tansy (2006) noted possible harmful effects of lampricides against a water beetle, for example.

Until recently, the major control method for the Australian plague locust (*Chortoicetes terminifera*) was aerial spraying with the pesticide fenitrothion over vast areas of inland eastern Australia to target hopper bands and adults. Inevitably, large amounts of non-target insects are killed by such measures, and impacts on species of possible conservation interest is undocumented. In North America, such broadcast of insecticides against pest Orthoptera has been reported as a threat to the Dakota skipper butterfly (*Hesperia dacotae*), and has been implicated as a cause of loss of small populations. In a similar context, aerial spraying of exotic weeds such as leafy spurge has also eliminated native forbs that are important nectar sources for the adult skippers, with the consequence that herbicide use for weed and brush control on private lands can be a principal threat to this skipper (USFWS 2005).

Likewise, the use of pesticides (including insecticides, molluscicides and herbicides) in water bodies may be a threat to some species. Wherever possible, taking steps to obviate any possible threat likely to arise is simply common sense – but in many cases prior knowledge of such measures may not be available locally. An adjunct to any insect management programme should be to conscientiously anticipate any such incidence. For example, insects on roadside sites may be within areas to be sprayed for weed control by local municipal authorities. Effective communication with any people or organisations whose cooperation would help to avoid such accidents is vital.

A more specific non-target case arises with parasiticides used for domestic stock. Avermectins (based on the microbial organism *Streptomyces avermitilis*) given to cattle, for example, remain in the animal faeces, and have been implicated in killing dung-breeding beetles and flies.

Eradication of exotic organisms by poisoning (above) also raises issues of non-target effects. These have only rarely been evaluated in specific cases, one being for mortality or sublethal effects on a weta (*Hemideina crassidens*) in New Zealand in the context of rodent control (Fisher *et al.* 2007). Captive individuals of this tree weta were exposed to the rat control baits containing the anticoagulant diphacinone, and analysed after periods of up to 64 days' exposure, to determine concentration of diphacinone residues in their bodies, as well as being observed for any changes in feeding behaviour, survival and body mass. Weta had detectable diphacinone, but did not accumulate it, and no adverse effects were reported. Safety of weta, together with many other New Zealand invertebrates, is of considerable concern in view of their attraction to anticoagulant baits, used widely for rodent control, in the field. Wider ramifications, including possible toxic effects on animals that eat weta, remain unclear.

The above contexts simply exemplify the wide range of considerations that may arise over pesticide and other chemical pollutant effects and that may need to be anticipated or considered in planning effective conservation in areas where such chemicals may occur. Excesssive use of washing detergent for laundering clothes in streams has been correlated with local loss of endemic dragonfly species on Mayotte (Comoro archipelago) (Samways 2003).

Overcollecting

Overcollecting is a particularly emotive topic in insect conservation, and is discussed here to provide background to non-entomologists on some problems relevant to conservation practice, and which can cause heated debate and mistrust. The perceived problems draw on (1) the reality that collecting of butterflies and some others has long been a popular hobby and is one of the major ways in which information of great value in conservation has been accumulated, and (2) that such collecting is commonly outlawed once species are listed for protection, through some form of legally ordained 'prohibition of take'. For many insects, collecting is indeed necessary, simply to identify the species clearly from study of voucher specimens. Unlike many better-known vertebrates, most groups of insects can not be identified by sight alone, and close examination (including dissection) may be necessary to differentiate many taxa. In short, assessing distributional or abundance trends often cannot be undertaken without capture and detailed study of voucher specimens of insects to confirm their identity.

The values of collecting insects may at times be opposed by arguments from a strong anti-collecting lobby, with the arguments for both sides summarised by Stubbs (1985) for Britain. Two major arguments are raised commonly in relation to the collecting of specimens by hobbyists. First, that such collecting *per se* is very rarely a threat to an insect, and is almost always subsidiary to other more prevalent threats such as habitat loss. Second, that prevention of collecting insects in those cases where it is not a demonstrable or likely threat deters the interest of the very people whose goodwill and support is necessary to help accumulate the knowledge needed in insect conservation. In such contexts, collecting is commonly still undertaken illegally, but the activities become clandestine, and the information gained on distribution and biology remains underground. Such concerns were expressed specifically for the localised Australian mangrove-frequenting lycaenid *Acrodipsas illidgei* by Beale (1998), for which alienation of hobbyist interests can indeed impede conservation progress. As Henning (2001) noted for South Africa 'it is usually only the butterfly collectors who, in the first place, became aware of the rarity of a species and it is often through them that the appropriate authorities are advised'. The long-term records of British butterflies that have yielded so much invaluable information on distributional changes (Asher *et al.* 2001), for example, have been the cumulative outcome of massive collector enthusiasm, interest and activity. In that well-documented fauna, collecting is no longer needed in order to identify butterfly species, other than for temporary capture of some small skippers in order to check antennal colour, and cases can indeed be made for collecting to become minimal as a possible threat in some local colonies. Nevertheless, even in Britain, collections of some other insect groups are necessary for identifications and other documentation to be backed by voucher material. The well-intentioned prohibition of take, so widespread in Europe (Collins 1987) and that has subsequently become a conservation exemplar elsewhere, can in practice become detrimental to basic conservation documentation and progress.

Kudrna (1986) cited examples of overcollecting of butterflies in Europe, and suggested that rare endemic species (which are particularly sought by collectors) and 'panoramic' species (those with distinct aesthetic appeal, usually conspicuous medium to large species, some of them rare) may be particularly at risk. One of the more notorious cases of reputed involvement of overcollecting in contributing to extinction of an insect is of the large copper butterfly (*Lycaena dispar*) in Britain in the mid-nineteenth century. The species was never common, and was confined largely to the fens of eastern England. Consensus exists that its decline was caused primarily by habitat loss, so that the draining of fenlands led to progressive fragmentation of habitat, allowing *L. dispar* to persist only in small isolated populations. In this condition, its extinction may have been accelerated by random fluctuations and/or overcollection, but Pullin *et al.* (1995) regarded these as secondary causes. In the United States, collecting was considered a threat to Mitchell's satyr (*Neonympha mitchellii mitchellii*), with strong commercial exploitation suspected to occur (USFWS 1997).

Concerns in Europe arose from the large number of collectors and activities of dealers exporting butterflies, particularly of rare species, which, like any other collectable object, are commercially desirable. In Australia, and many other places, intensity of collecting is likely to be rather low in comparison with the former European scene, and undertaken by relatively small numbers of enthusiasts, whose collective impact is likely to be minimal. In very extreme cases – such as species known from single small populations on single or few sites – any collecting might, of course, 'tip the balance' and be undesirable, but such circumstances appear to be very rare. A survey of Australia's butterfly enthusiasts (Greenslade 1999) revealed their many concerns, and Kitching (1999) considered that harm from the likely modest activities of butterfly collectors in Australia would be by far outweighed by the benefits of increased knowledge from their activities. Deeper concerns can arise when collecting is associated with habitat destruction, such as large scale bark stripping or digging of ant nests while seeking early stages of elusive or desirable Lycaenidae. Destruction of trees to collect commercially desirable beetles associated with timber is a major concern in parts of the Old World tropics (New 2005b). Such issues are addressed in a number of 'codes for collectors' designed in various parts of the world. It is inevitable that there will always be a few greedy or commercially motivated collectors, some of whom may go to enormous lengths to collect the rarest species for the 'black market'. Legislation does little to deter such activities, other than by largely accidental detection of offenders.

In general, moderation and responsibility should underpin specimen collecting, and be accompanied by the reality that many groups of insects, even if species are signalled for conservation value, are only rarely the focus of any non-scientific collecting. In order to satisfy the generally

Box 5.2 · *Codes for insect collecting, and their conservation roles*

Suggestions of the harmful effects of overcollecting, particularly of more 'popular' or commercially desirable insects such as Lepidoptera and larger Coleoptera, have led to a number of 'Codes for Collectors' or 'collecting policies' to encourage appreciation of the need for responsibility and to provide practical advice on how to avoid harm to the taxa involved and their habitats. They seek to make hobbyists and others aware of the needs for restraint and care, and most such documents have arisen directly from groups of entomologists, some from people concerned over the wider ramifications of 'collecting bans' as a measure to protect the species. The codes have considerable common elements, in emphasising ethical and practical responsibility in the collecting process and in care and documentation of specimens.

An early British code (JCCBI 1971) was an important fore-runner in setting an agenda for later codes, and was stimulated by fears that collecting efforts would become increasingly detrimental as losses of habitat increased. That code provided a series of pointers under six main headings (Table 5.2.1), and these have been paralleled or emulated in several influential later documents, such as that by the Lepidopterists' Society (1982). The various considerations involved may include points of direct concern to any species conservation programme where 'take' of specimens for any purpose is contemplated. Particular wording, at least, may need to be modified for local conditions or to comply with local regulations or laws.

Table 5.2.1 *Some issues listed by JCCBI (1971) in* A code for insect collecting

1. Collecting: general
   1.1 No more specimens than are strictly required for any purpose should be killed.
   1.2 Readily identified insects should not be killed if the object is to 'look them over' for aberrations or other purposes: insects should be examined while alive and then released where they were captured.
   1.3 The same species should not be taken in numbers year after year from the same locality.
   1.4 Supposed or actual predators and parasites of insects should not be destroyed.
   1.5 When collecting leaf-mines, gas and seed heads, never collect all that can be found: leave as many as possible to allow the population to recover.

1.6 Consideration should be given to photography as an alternative to collecting, particularly in the case of butterflies.

1.7 Specimens for exchange, or disposal to other collectors, should be taken sparingly or not at all.

1.8 For commercial purposes insects should be either bred or obtained from old collections. Insect specimens should not be used for the manufacture of 'jewellery'.

2. Collecting: rare and endangered species

   2.1 Specimens of 'listed species' (in this case, of macrolepidoptera listed by JCCBI) should be collected with the greatest restraint. As a guide, the Committee suggest that a pair of specimens is sufficient, but that those species in greatest danger should not be collected at all. The list may be amended from time to time if this proves to be necessary.

   2.2 Specimens of local distinct forms (here, of macrolepidoptera, particularly butterflies) should likewise be collected with restraint.

   2.3 Collectors should attempt to break new ground rather than collect a local or rare species from a well-known and perhaps over-worked locality.

   2.4 Previously unknown localities for a rare species should be brought to the attention of this Committee, which undertakes to inform other organisations as appropriate and only in the interests of conservation.

3. Collecting: lights and light-traps

   3.1 The 'catch' at light, particularly in a trap, should not be killed casually for subsequent examination.

   3.2 Light-trapping, for instance in traps filled with egg-tray material, is the preferred method of collecting. Anaesthetics are harmful, and should not be used.

   3.3 After examination of the catch the insects should be kept in cool, shady conditions and released away from the trap site at dusk. If this is not possible, the insects should be released in long grass or other cover and not on lawns or bare surfaces.

   3.4 Unwanted insects should not be fed to fish or insectivorous birds and mammals.

   3.5 If a trap used for scientific purposes is found to be catching rare or local species unnecessarily it should be re-sited.

   3.6 Traps and lights should be sited with care so as not to annoy neighbours or cause confusion.

4. Collecting: permission and conditions

   4.1 Always seek permission from the landowner or occupier when collecting on private ground.

   4.2 Always comply with any conditions laid down by the granting of permission to collect.

   4.3 When collecting on nature reserves, or sites of known interest to conservationists, supply a list of species collected to the appropriate authority.

   4.4 When collecting on nature reserves it is particularly important to observe the code suggested in section 5.

5. Collecting: damage to the environment
    5.1 Do as little damage to the environment as possible. Remember the interests of other naturalists: be careful of nesting birds and vegetation, particularly rare plants.
    5.2 When beating for lepidopterous larvae or other insects, never thrash trees and bushes so that their foliage and twigs are removed. A sharp jarring of branches is both less damaging and more effective.
    5.3 Coleopterists and others working dead timber should replace removed bark and worked material to the best of their ability. Not all dead wood in a locality should be worked.
    5.4 Overturned stones and logs should be replaced in their original positions.
    5.5 Water weed and moss which has been worked for insects should be replaced in its appropriate habitat. Plant material in litter heaps should be replaced and not scattered about.
    5.6 Twigs, small branches and foliage required as food plants or because they are galled, e.g. by clearwings, should be removed neatly with secateurs or scissors and not broken off.
    5.7 'Sugar' should not be applied so that it renders tree-trunks and other vegetation unnecessarily unsightly.
    5.8 Exercise particular care when working for rare species, e.g. by searching for larvae rather then beating for them.
    5.9 Remember the Country Code!
6. Breeding
    6.1 Breeding from a fertilised female or pairing in captivity is preferable to taking a series of specimens in the field.
    6.2 Never collect more larvae or other livestock than can be supported by the available supply of food plant.
    6.3 Unwanted insects that have been reared should be released in the original locality, not just anywhere.
    6.4 Before attempting to establish new populations or 'reinforce' existing ones, please consult this Committee.

modest requirements of collectors, Sands and New (2002) suggested a number of measures that might serve to reduce tensions between collectors and conservation managers in Australia, and allow fuller and welcome exchange of information and experience of species of conservation concern. Such steps may need modifications to regulations or law in places but included considerations of (1) whether particular protected insects may be available for limited collection on some sites with large viable populations, while maintaining total protection of demonstrably more vulnerable populations; (2) organisation of 'open days' at which managers and hobbyists might meet on sites to exchange information and perspective; and (3) actively encouraging collectors to participate in the extensive surveys (including those for protected areas such as national parks) needed to augment knowledge of distributions and biology. These, and other means, merit consideration in other countries where foundation knowledge of insect groups for conservation is lacking and where it is acknowledged that such cooperation and goodwill is valuable. The longer term transformation involves, in Kudrna's (1986) words, 'neutralising all harmful aspects of collecting' and giving the collector 'a brand new, respectable, image'.

Special constraints on collecting may be needed in specific contexts – for example, to accompany phases of recovery plans. For example, if a reintroduction has involved translocation or other release of individuals in a new site, removal of any individuals by collecting might be unwelcome and destroy the programme.

Most insects are dispersed reasonably widely in their habitats, with the implication that collecting requires increasing effort as the targets become scarcer. The problem is paralleled in detecting and monitoring species for conservation, of course. Their habitations may also be remote. Thus, *Colophon* stag beetles in South Africa are threatened by overcollection in montane areas likely also to be affected by climate change, for commercial sale. All 17 species live on the summits of mountain ranges (Geertsema & Owen 2007), and the only practical protection (other than legislation to prohibit collecting) has been their remoteness and general inaccessibility. This is not wholly satisfactory, because that very remoteness also renders these sites impossible to patrol effectively. In this case, determined beetle-poachers may remain undetected and continue to obtain beetles.

In contrast to the common distributional pattern of an insect occurring in very low density over its range, in some other contexts, particular insects can aggregate naturally in large numbers, or be attracted in large numbers to baits of various sorts. Under such circumstances they can present a 'bigger target' for unwitting or deliberate harm, and some comment on the conservation implications of potentially heightened vulnerability is needed here. Particular sites, such as some hilltops where rare species may predictably gather to seek mates (p. 96) may be particularly vulnerable as 'classic' collecting localities. If such sites are remote, a visit to them may be a major exercise and expense, and collectors may then be tempted to take specimens excess to their own requirements, for exchange or 'for a friend'. Even when a species may seem abundant on a hilltop, the population observed on that small area may represent a high proportion of individuals normally dispersed over a radius of several kilometres or more.

A few insect species have been noted for their spectacular vast seasonal aggregations, which regularly excite media comment and wide interest, sometimes associated with local tourism operations. In such phenomena, much or all of a population (or regional representation of a species) gathers in one place or a limited number of sites to hibernate or aggregate as a regular feature of their seasonal life cycle. By far the best-known such species is the wanderer or monarch butterfly, *Danaus plexippus*, in North America and Mexico. Adults move in autumn to particular sites in California or Mexico, depending on whether they originate from the western or eastern regions of North America, over which the species becomes widely distributed during spring and summer. They overwinter in these sites, clinging to conifer trees. The conservation of this species depends heavily on protection of the forest groves used traditionally by the butterflies, where the whole continental population of the species is then concentrated, from human disturbance and destruction by forestry or more casual cutting of trees. Even limited forest thinning in Mexico increases butterfly mortality by changing the local microclimate (Brower 1996).

A less charismatic case of mass aggregations involves the Bogong moth, *Agrotis infusa*, in southeastern Australia (p. 78). Adult moths move to high ground in the alpine region in early summer to aestivate in large aggregations in caves and under rocky overhangs. In the past *A. infusa* was used by local people for food, but it is now most remarked not for this remarkable phenomenon, but for the putative nuisance it causes during its spring migration, when it is attracted to venues such as sports grounds by lights in vast numbers. It is not threatened by collectors, but its conservation may need conscious attention to its public image! The moths may also have wider conservation relevance, in that they may be a conduit for arsenate chemicals to remote areas from the agricultural lowlands from which they move. Again, the moth's wellbeing depends on the availability of specific topographic sites, in this instance of alpine refuges.

The Jersey tiger moth, *Panaxia quadripunctaria*, on the Island of Rhodes (Greece) is a striking example of the tensions between tourism and insect conservation. The moth reaches exceptionally high population levels in the so-called 'Valley of the Butterflies', which has consequently become a major tourist drawcard on the island (Petanidou *et al.* 1991), in association with the only known natural forest of *Liquidambar orientalis* in Europe. Tourist activities have been implicated directly in decline of the moth, through direct disturbance to resting moths and trampling and destruction of vegetation. Collection is also implicated, with Petanidou *et al.* (1991) estimating that conservatively 100 000 insects per generation (year) were removed by tourists. Conservation measures then included increasing information to visitors, wider environmental education programmes, increased guarding of the area, prohibition of activities such as clapping or whistling (which cause moths to become restless), not allowing people to disperse from marked paths (for example, by fencing the pathways), and general revegetation of the Valley area by planting of *Liquidambar* trees, which were also protected from grazing stock such as goats by wire surrounds. For this species, the conservation measures taken were clearly needed, low cost, rapid to execute, and − it seems − largely effective and also harmonious with sustaining its use in fostering tourist interest.

A second relevant context is that certain insects may be attracted in vast numbers to lights or chemical baits. Indeed, using 'light traps' for moths is a very common method used by collectors to obtain specimens, and also an important survey method. Codes of conduct imply that killing agents should not be used in such general attractant traps, and that unwanted captures should be released alive and unharmed. In doing so, care must be taken to ensure that their vulnerability to predators (perhaps, in particular, birds) is not increased. For example, many moths are wholly crepuscular or nocturnal and, should they be discarded from the enclosed trap in early morning, may not be able to disperse and hide. It is generally recommended that the moths be retained in the shelter of the trap container until the following evening and then allowed to disperse naturally. Street lights have been implicated in the decline of a variety of insects attracted to them, when the insects become more conspicuous and vulnerable to bats and birds.

In general, any insect collecting or sampling method in which the non-selected catch is killed or harmed should not be employed uncritically in areas where known threatened or notable species occur. Malaise traps, for example, readily catch a wide variety of actively flying insects, and species of conservation concern may be captured by accident. On the other hand, such traps (as with pitfall traps for Coleoptera) have yielded a number of notable records of such species. However, their use on sites where insects of current conservation interest occur is generally unwise and, even, provocative. Should such species be found unexpectedly during other ecological exercises involving such traps, the trapping programme should be modified for capturing insects alive, or stopped.

For insect species (such as certain rare butterflies and beetles in various parts of the world) which are genuinely threatened, or genuinely believed to be threatened, by commercial desires and overcollecting, the management options might include consideration of higher profile notice of protection need. A step such as promoting for listing under CITES serves to widen awareness of the possible plight of such species and, notionally, to gain information, on the numbers of specimens in detected trade through a permit system. Two practical problems caution against widespread dependence on such steps in more immediate practical contexts. First, insects are small and easily transported clandestinely by determined individuals. Second, many insect species are very similar in appearance, and most people policing quarantine stations and the like are not entomologists. It is not reasonable to expect them to be able to differentiate between taxa that many a specialist would have difficulty separating. This general problem arises from prohibitions on take of certain species whereas numerous closely similar and related species may not have any need for equivalent protection. One 'solution' is that adopted for birdwing butterflies under CITES Appendix 2, namely to list 'all species of birdwings' as a precautionary measure so that the few species genuinely threatened by overcollecting are included with much more common and secure species. Although the various species of birdwing are sometimes difficult to tell apart, the general appearance of 'a birdwing' is very characteristic and easily recognisable by non-expert inspectors. Understandably, measures such as this have led to critical comments from hobbyists, foisted with a sometimes complex permit-gaining exercise in order to obtain even common species for their collections.

A similar 'umbrella step', with parallel motivation of protecting look-alike species, was enacted in Western Australia in the 1970s, when 'all species of jewel beetle' (Buprestidae) were listed for protection under State law. Some European legislations have also included blanket protection in prohibiting collection of all butterflies, or a similar phrasing (Collins 1987).

Any regulatory prohibition of collecting insects without parallel protection for the habitat of the species involved is unsatisfactory: such regulation is not in itself effective conservation management. As Collins (1987) remarked, in a sentiment echoed many times since, 'conservation [of Europe's insects] will depend upon protection and appropriate management of vulnerable habitats coupled with judicious control of direct species exploitation.' As noted above, bland 'prohibition of collecting' is a decidedly two-edged sword in many insect conservation programmes. More broadly, any case made for total prohibition of collecting any insect species needs very careful and informed decision in relation to threat versus information benefit, and to determine whether prohibition (rather than strict regulation) is warranted on grounds of threat. Constructive alternatives may at times be available. It may, for example, be possible to encourage responsible surveys for species, with permission to retain vouchers in private collections, or to allow collectors limited access to selected populations or sites. In Victoria and Queensland, the State conservation agencies have issued permits to the State Entomological Societies, whereby society members may explore for, and collect limited numbers of, listed threatened insect species (unless excluded specifically on grounds of 'real threat'), with the data to be contributed to the central pool of knowledge on that species. This step has helped to provide additional distribution records for a number of species of high conservation interest.

Some forms of potential overexploitation of insects may not initially be obvious, but may occasionally need attention. Thus, larvae of dragonflies are collected and sold commercially for fishing bait (as 'mudeyes') in Australia, and there may be some potential to cause harm to rare species at some sites, simply because of their indiscriminate inclusion in multispecies catches.

The themes noted in this chapter may or may not intrude on an insect management plan. However, neglect of their possible incidence, likely impacts, and means to abate them may prove catastrophic. 'Threat assessment' is the foundation of practical conservation management, as emphasised in Chapter 1. For most insects, one or more threats (most commonly associated with habitat loss or degradation) may be obvious, but it is important also to ensure that no major threat is overlooked. Two possible, complementary approaches to this are indicated in Boxes 5.3 and 5.4. One involves considering the possible threats to the habitat or site, as a means to compensate partly for lack of detailed knowledge of threats to the focal species, and considering the relevance of each of these to that species. The second is to compile a list of threats recorded for members of the relevant insect order, family or guild, as a guide to what has been suggested or proved to be important in related situations and, again, appraising the relevance of each to the species now of interest.

Box 5.3 · *Defining a threat portfolio for an insect: precaution and responsibility. Hungerford's crawling water beetle in the United States*

Hungerford's crawling water beetle, *Brychius hungerfordi* (Haliplidae), occurs in well-aerated riffles of a few small slightly alkaline streams in Michigan (USA) and Ontario (Canada). The reasons for this very restricted distribution are not understood and there is a dearth of historical information to determine whether the beetle was formerly more widespread. Larvae tend to occur on dense aquatic vegetation (on which they feed) along stream edges, and adults are found in more exposed water where they feed on algae on rocks and stones. Detailed knowledge of the beetle's habitat requirements is very inadequate, and because threats can largely only be hypothesised, Tansy (2006) adopted a precautionary approach to threat evaluation, as implied in the statement '. . . in the absence of data on threats to *B. hungerfordi*, we are proposing these possible threats through inference based on information available on impacts to the habitat in which the species is found or impacts to other aquatic invertebrates.'

The following topics associated with habitat destruction and modification were signalled:

Stream modification, thought to be the primary threat, may include either or both of (i) physical destruction of the stream habitat and (ii) degradation of water quality. Several specific aspects were noted:

(a) Removal of beaver dams. In some sites beaver impoundments appear to be important in maintaining suitable habitats, with high beetle densities immediately below dams because of riffles and highly aerated water. Removal of dams could lead to local extinctions. In addition, new beaver dams might eliminate known suitable habitats by flooding, so that either positive or negative effects might ensue.

(b) Road crossings and culverts, if poorly designed, can lead to sedimentation of the streams, and cleaning of culverts and adjacent areas may need to be undertaken carefully to avoid lowering water quality. Culverts may also be a barrier to beetle dispersal, and also a source of pollutant entry from road run-off. Associated roadwork or bridgework may cause local disturbances.

(c) Projects such as bank stabilisation may have both beneficial and harmful effects. They may reduce erosion and sedimentation, but use of an impervious cover to moist soil near the waterline may eliminate potential pupation sites for the beetle.

(d) Logging in the riparian zone, dredging for channelisation, and other sporadic activities are possible threats to habitats, and may need to be assessed individually if inhabited streams are targeted.

Other possible threats noted include stocking of steams with insectivorous fish for recreational activities, leading to a possibility of increased predation over 'normal' (but for *B. hungerfordi*, unknown) levels. Activities related to fish management (such as removal of dams and culverts to facilitate fish movements or, conversely, construction of fish ladders) may also have adverse effects. Electrofishing and disturbances (trampling) by anglers could also be possible threats.

Box 5.4 · *Projecting a threat portfolio for an insect species: lessons from a higher taxonomic group*

To complement the process of hypothesising threats to a particular insect species from a study of its immediate environment (Box 5.3), it may be feasible to compile a list of the various threats recorded tangibly for the group of insects to which the focal species belongs, to provide an encompassing 'checklist' for evaluation for their relevance to any individual species. Thus, Chelmick *et al.* (1980) listed the various anthropogenic habitat changes that might influence British Odonata, as below, and this list emphasises the dual considerations of the aquatic environment (for larvae) and nearby areas (for adults). Many dragonflies and damselflies subsequently studied for conservation have manifested declines attributed to one or more of these factors. Even in cases where one or more threats to an insect are obvious, a wider perspective (such as from a checklist for the particular insect group paralleling the one below) is invaluable in ensuring that no important topics have been neglected.

The factors listed by Chelmick *et al.* (1980) are:

1. Loss of water bodies (breeding habitat).
2. Modification of ditches and rivers, such as influencing water levels, scooping out vegetation, creating sharp margins, or altering slopes of banks.
3. Drainage.

4. Making fluctuating water levels by pumped drainage schemes and reservoirs.
5. Use of chemicals against mosquitoes and wind drift of terrestrial insecticides.
6. Use of herbicides in weed clearance, including those applied against margin vegetation.
7. Pollution, including run-off of agricultural chemicals.
8. Over-management by amenity or fishing interests.
9. Overstocking with fish or ducks, which may feed on aquatic larvae; excessive droppings from birds can lead to eutrophication of water bodies.
10. Loss of immediate natural surroundings, including shelter belts.
11. Loss or modification of wider hinterland.
12. Lack of management, resulting in shading or choking of water with silt and plants.
13. Afforestation leading to excessive shade and lowering of water bodies.

Summary

1. Threats to insects extend well beyond the habitat changes discussed earlier, and some additional ones are noted and discussed in this chapter. Any or all of these may occur in any particular insect conservation case.
2. Alien species (many of them exotic) of plants and animals pose many threats to native insects by displacing them or otherwise interfering with supply of their critical resources. As examples, specific food plants may be outcompeted by alien aggressive weeds, or mutualistic ants displaced by 'tramp' species. Many such species and their effects are not immediately obvious, and they may be difficult to detect, monitor and evaluate.
3. Detection and eradication of alien species is an important component of insect species conservation, with a wide variety of gambits available. The various options for suppression or eradication may themselves need careful evaluation to select those least harmful to other, desirable, attributes of the habitat or to the focal species itself.
4. Pesticide effects or, more widely, chemical pollutant effects, may also occur either by accident or as a byproduct of crop or other commodity protection operations within the vicinity of a conservation focus species.

5. Overexploitation, most notoriously for insects by 'overcollecting' of commercially desirable taxa such as rare butterflies and beetles, is a highly emotive topic in insect conservation. It is apparently only rarely a genuine threat to particular species. Legal 'prohibition of take' is a common outcome of listing insects for protection and conservation priority, but the balance between garnering valuable biological and distributional information on those species from hobbyists and non-professional entomologists (which is a significant or predominant component of the information available on many species) and alienation of those interests must be considered very carefully. Collection/exploitation must be undertaken responsibly, and steps to increase cooperation between all parties interested in the species' conservation pursued whenever possible tensions are predicted to arise.

6 · *Adaptive management options: habitat re-creation*

Introduction: improving habitats for insects

The discussions in previous chapters provide some of the perspective underpinning what might be considered 'good management', by helping to define clear objectives of a conservation programme and the actions needed to accomplish them, either on particular sites or more widely. Considerations of habitat extent and quality are integral to this, with enhancement (sometimes, reinforcement) and restoration of sites and resources the central issues for both short term and longer term conservation. In addition to the intrinsic features of any single site, the place of that site within the wider landscape must be considered in relation to implications of isolation, with the maintenance or restoration of connectivity between populations (through sites) sought at almost every opportunity. Agricultural landscapes, for example, are a prime focus for conservation of some insects, and may present very complex mosaics of natural remnant patches across and within a wide variety of highly modified habitat and resource conditions. They provide many opportunities for manipulation for conservation of individual species (New 2005a; Ouin *et al.* 2004), because many of their structural and compositional features may be modifiable without compromising their primary purpose of providing commodities for humanity. Indeed, emphasising the roles of insects in promoting ecosystem processes seen as beneficial by landowners may facilitate conservation measures for species of lesser direct interest to them.

Particularly in Europe, so-called 'brownfield sites' have attracted considerable interest for insect conservation, and have also provided many opportunities for restoration. These are sites used previously for development but now abandoned. They range from old quarries and factory sites to former railways and roads, but are broadly alike in that they commonly have nutrient-poor soils and these may provide ideal opportunity for colonisation by local flora not well adapted to the more complex communities of grasses and herbs found on richer soils. Brownfield sites may thereby provide successional habitats that may otherwise not be strongly represented locally. Specific steps may be available to 'tailor' such sites for particular notable insects. Strauss and Biedermann (2005) drew attention to the importance of brownfield sites for endangered leafhoppers (Homoptera) in Germany, and similar examples could be multiplied extensively. For the leafhoppers, age of site (that is, time since abandonment) was the most important determinant of species occurrence, emphasising again the dependence of many specialist species on particular successional stages, and the need to provide for their continuity as a major component of habitat management.

Habitat or biotope 'improvement' is a potentially universal focus in insect species conservation, as the main means to assure the resources needed by those species.

Restoration

Many endangered insects depend on some form of habitat restoration for their security, as a direct consequence of their former high quality habitat having been lost or degraded to levels where resources are insufficient to sustain the species or population at its former level. Restoration depends on knowing the causes of degradation, and these may sometimes be from unexpected quarters. Thus, degradation of littoral vegetation and banks of the River Tisza in Hungary by recreational fishermen was associated with decrease in numbers of several rare species of Odonata (Müller *et al.* 2003). Suggested remedial management in that case involved (1) limiting the number of permanent fishing stands on the river banks and (2) spacing these out to separate them clearly by stands of undisturbed vegetation – so that a key to preventing further habitat loss was to regulate use of the river banks by anglers, in order to conserve riparian vegetation. This case applied to a number of coexisting species of Odonata at these sites. Any restoration directed toward a particular species depends on ecological understanding of that species (Pavlik 1996), even though some general principles of habitat restoration for insects have started to emerge.

Three levels of management activity may need to be considered:

1. Restoration of sites, either single sites or multiple sites across a landscape.
2. Restoration of specific resources, either generally or to particular levels.

3. Restoration of populations, either by augmentation of existing populations or establishment of new ones.

Morris *et al.* (1994) used the term 'ecological engineering' in relation to habitat restoration for insects, and the above are simply the major dimensions of this. All these approaches require protection of what is there already, and avoiding any further loss of space or needs. These may at times involve activities that seem tangential to direct focus on the species involved. For example, prohibition of collecting fallen wood as firewood along local roadsides was a management recommendation for the small ant-blue butterfly (*Acrodipsas myrmecophila*) in Victoria, because the rare ant with which its caterpillars obligately associate (and on whose larvae they feed) nests in dead wood, which was scarce in the area owing to its continued removal. In this case, dead wood was a critical resource for the ant mutualist. Elsewhere, its values for insects are recognised widely, as a substrate or resource for numerous species.

Particularly in Europe (Speight 1989), conservation of saproxylic insects has attracted considerable attention, but the importance of dead wood for insects in tropical forests is also appreciated widely (Grove & Stork 1999).

Fortuitously, a number of insect conservation programmes overlap in that the species occupy the same individual sites as species of vertebrates or flora also targeted for conservation, so that these wider considerations may necessitate integrating the insect conservation optima with others for greatest collective benefit. However, beyond assuring basic security of the sites involved, much of the management need will usually extend beyond generalities and must be tailored to the individual species involved. Enhancement of particular plant species as larval or other food plants may be an obvious need, but the extent of this, the methods involved in propagating or transplanting the species needed, and features such as the optimal age structure of plant stands needed may be entirely unknown. Should the plant itself be endangered (and then possibly a correlate of the insect's threat status), some botanical knowledge may well be available from other sources, but supplies for experimentation will often be very limited. Threatened or localised herbivorous insects are often associated with threatened or localised food plants.

The management of a critical plant resource based on effective ecological understanding of the insect of concern is demonstrated by the specific recommendations for the large copper butterfly, *Lycaena dispar* (p. 131), in Britain, and the reasoning behind these measures:

1. Maintain food plants (the great water dock, *Rumex hydrolapathum*) in open sunny positions by use of cattle grazing and/or biannual mowing. Rationale: female butterflies prefer to lay their eggs on such exposed plants, and the open position may result in reduced predation of pre-hibernation stages by invertebrates.
2. Plants should be growing actively at the time of oviposition. Rationale: they then provide the most nutritious food for young larvae, enabling them to accumulate sufficient reserves to overwinter.
3. Plants should not be in especially low-lying positions. Rationale: such sites are affected by flooding, also a cause of caterpillar mortality.
4. Large areas of open fen should be maintained. Rationale: the larger area provides the arena for male territories and assures adequate nectar resources for adult butterflies.
5. Maintain a network of sites meeting these requirements. Rationale: provides sites for colonisation and a buffer against local extinctions, and allows for the normal dispersive behaviour of the butterfly (Pullin *et al.* 1995).

Formulating such specific recommendations from a basis of sound knowledge and clear purpose is the aim of many insect conservation programmes. However, for most insects, an informed level of focus equivalent to the above case is initially difficult or impossible. As knowledge accumulates, conservation measures can become increasingly precise in their aims. The ability to incorporate progressive refinements is needed in any sound conservation plan.

Many insect management activities and programmes have focused on and been formulated for single sites, but in many cases methods trialled or proven for one site may not be transferred uncritically to others – even though the central activity is part of a valuable general protocol for the species. Knowledge of threshold values or conditions for critical resources or conditions may be very pertinent. For example, O'Dwyer and Attiwill (2000) recommended a target of 40% ground cover of *Austrodanthonia* grasses to be restored on a grassland site for golden sun-moth in Victoria, but later studies suggest that this level may not always be necessary to sustain populations elsewhere. Likewise, some knowledge of host plant age and condition preferences facilitates estimation of what may be needed. Caterpillars of the skipper butterfly *Hesperilla flavescens* (p. 38) in southeastern Australia are monophagous on the sedge *Gahnia filum* on near-coastal sedgelands, but only young foliage is suitable as food and attractive for oviposition. Loss of sedge habitat and lack of natural regeneration have dictated need for management to replenish sedges and assure food supply. Both in South Australia (for *H. f. flavia*: Coleman & Coleman 2000) and Victoria (for *H. f. flavescens*: Savage 2002), this key aspect of management has been pursued by burning old sedge tussocks to promote regeneration, translocating young sedges, planting of nursery-grown stocks, and seed-scattering, representing a variety of available options for restoring a supply of the sedge. Unusually, burning management for the skipper can be undertaken at very fine scales, because even the individual tussock or small groups of tussocks can be burned, with adequate precautions, and those tussocks can be selected (on the basis of absence of the conspicuous larval shelters of foliage) as unoccupied by caterpillars. Fresh growth occurs from burned tussocks within a few weeks, and is attractive to female butterflies for oviposition.

For any species, field trials may be needed to assess the 'best' method(s) to employ. In augmenting lupins for Fender's blue (below), more than 12 000 seeds were sown and 600 seedlings transplanted. Fewer than 10% of seed sown in 1996 germinated or survived until 1997, and most of the transplanted seedlings had died by 1998. Fewer than 60 new plants were evident by 1999 (Schultz 2001).

Much habitat restoration for terrestrial insects centres on or otherwise involves plants – either enhancement of desirable ones, such as food plants, or elimination of undesirable ones such as aggressive weeds, or both of these together. Both are among the most frequent activities suggested for insect conservation management. Simple extension of habitat area may be possible: Sands (1999) suggested that the range of *O. knightorum* (p. 78) might be extended by planting of *Alexfloydia* grass in suitably protected sites near or within the butterfly's restricted range, followed by introduction of the insect. However, habitat restoration may involve far more than just augmenting a single resource such as a food plant, and the integration of various management aspects needs careful thought. Each aspect of restoration may need to be integrated with others and considered for site and wider population or landscape benefits. At either scale, but with emphasis here on single-site management, some form of quantitative or semi-quantitative target should ideally be available – either in terms of process ('planting X individuals of plant A') or outcome ('augmenting the site to carry a population of Y adult insects'), each to be targeted over a specified period, and based on sound knowledge of resource requirements. Thus, for Fender's blue, above, the three major requirements of a habitat are larval host plant (the lupin *Lupinus sulphureus kincaidii*), native forbs as adult nectar sources, and grasses and forbs that represent the historical short-grass structure of the prairie habitats (Schultz 2001). Schultz and Dlugosch (1999) provided an estimate that suitable habitats for adult Fender's blue should support approximately 20 mg sugar m^{-2} of nectar from native forbs and that for caterpillars should support approximately 40 leaves m^{-2} of *Lupinus*. This is one of very few insects for which such a precise estimate of satisfactory resource supply has been made, and such information is in any case difficult to obtain and almost always relies on correlative measurements. Relating precise thresholds to a given or anticipated population size or site carrying capacity remains unusual and difficult, but any general guidelines may be constructive in helping to assure that resources are sufficient and well above marginal levels of supply.

Box 6.1 · *Butterfly distribution, resources, and planning for habitat enhancement*

The very specialised needs of many threatened insects are only now starting to be understood, and many aspects of practical habitat enhancement are based on generalised principles, rather than on detailed direct knowledge of the species involved. Dover and Rowlingson (2005) studied a population of the western jewel butterfly (*Hypochrysops halyaetus*, Lycaenidae) on a small urban bushland reserve near Perth, Western Australia, where the locally endemic southern form of this species is 'vulnerable'. It has a single larval food plant (*Jacksonia sternbergiana*) (Fabaceae) and caterpillars associate with the localised ant *Camponotus perthensis*.

Dover used mark–release–recapture techniques to study aspects of dispersal, distribution, population size and habitat 'preferences', with one localised colony on a site only 80 m × 20 m within the reserve studied intensively over about a month. Within this small area, 1158 butterflies were marked individually (by a number on the underside of a hind wing). The site was divided into 96 small recording units to facilitate detailed study. It comprised several distinct areas differing in vegetation composition and physical features such as the amount of bare ground. Butterfly density decreased sharply on moving from this site into adjacent bushland.

Females tended to move further than males, although males tended to move faster. Butterflies appear to prefer degraded bush, essentially that representing a post-disturbance regime with high densities of *Jacksonia*. Male numbers correlated positively with the amount of bare ground, possibly related to their need for perching sites from which to seek mates. Females were negatively correlated with denser ground shrubs, particularly those that indicate more mature bushland vegetation. There were suggestions, by the northern aspect of the site, that the thermal ecology of *H. halyaetus* may be an important component in habitat selection.

Dover and Rowlingson (2005) inferred that any concentration on 'improving' vegetation condition might in fact lead to population declines, because of loss of bare ground or early successional vegetation. Fire management may be important in achieving the balance required. *J. sternbergiana* is a post-fire opportunist species, with botanical studies indicating that it needs high levels of phosphorus, as would be supplied normally from ash-beds after fires. If this is so, substitution of fine-scale mosaic burning management by mechanical removal of vegetation may not be adequate to maintain it. Steps suggested for experimental re-creation of *H. halyaetus* habitats without fire are to determine (1) whether high densities of *Jacksonia* can be created; (2) whether such areas are suitable for *Camponotus*; and (3) whether such areas will be colonised by the butterfly. The butterfly's lack of reliance on late successional vegetation also opens up the possibility for use of other degraded bush sites that normally would be considered of only minimal conservation value for insects and flora for future translocations.

Augmentation of resources such as plants is usually needed in some concentrated form over rather small areas, to enhance a site's carrying capacity for the focal insect species. It may occasionally be undertaken with wider benefits planned, or over much larger parts of a species' range. Thus, widely dispersing species of birdwing butterfly may be 'concentrated' by provision of the vine food plants normally also dispersed widely, as in the case for ranching such species in New Guinea (p. 178; Parsons 1992). Birdwings are adept at tracking such resources in the wide environment (Sands *et al.* 1997). In Queensland, planting of *Pararistolochia* vines as food for caterpillars of *Ornithoptera richmondia* is a major component of the species' recovery programme (Sands & Scott 2002). More than 32 000 nursery-grown vines had been sold to the public by one nursery, and schoolchildren in Queensland and New South Wales (the only states in which the butterfly occurs) are among the many people aiding in planting vines and monitoring the butterfly's spread in response to this, with many plantings being made in school grounds or private gardens as well as in more natural sites in remnant native forest. These plantings have gradually fulfilled the important functions of (1) providing corridors between habitat patches and (2) providing habitat for new colonies of *O. richmondia*. Plantings thereby have both local and wider landscape roles in the butterfly's conservation.

However, once critical resources are known, comparison of their extent on sites on which the target species thrives and those contemplated for restoration may indicate some level of need and help focus the aim and extent of the work needed. Often, such comparisons are not possible, simply because relatively pristine or undamaged sites are not available as a 'baseline' for comparison. Likewise, the optimal procedures for sustaining existing resources may be unclear.

Some management approaches

Operations such as burning or grazing to influence vegetation structure are cited frequently in insect habitat management, but these may need very careful planning and regulation. Any such dynamic management may involve consideration of 'tradeoffs' with negative short-term impacts balanced against the hope of achieving long-term success. Thus, as Schultz and Crone (1998) noted, fire can be an important tool to help control invasive weeds in grassland and to stimulate growth of native plants, but may destroy much of the existing invertebrate fauna. Much such management can be considered high risk. Any burning operation, for example, must be planned as rationally as possible to ensure that likely benefits are greater than the damage caused, and many variables of space and time may need to be considered in assessing this balance.

The small urban sites near Melbourne occupied by the Eltham copper butterfly are subject to weed invasion and canopy closure due to natural succession, and these are among the processes that render the sites increasingly inhospitable to *P. p. lucida*, as an example of the numerous insect species for which seral succession is a major threat to habitat suitability. Burning was considered in the mid 1990s as a management option to 'rejuvenate' sites, and two of the main sites were burned in 1998 (New *et al.* 2000). The background considerations in this example involved both social ones (damage to adjacent residential properties) and scientific ones of wide relevance, and which are paralleled in many other cases where 'fire management' may be contemplated. These considerations included (1) time of year to burn; (2) the intensity of the fire needed; (3) whether to burn all or part of the site or sites, and if the latter, which parts; and (4) that there was risk of destroying the entire resident population of the butterfly on the site(s). These and other factors were discussed extensively within the species' management group, with the following outcomes: (1) resolve that without drastic management intervention, the sites would become non-viable for the butterfly within a few years, so that the butterfly would become extinct there; (2) that burning was the most suitable option to pursue (the only alternative of labour-intensive hand weeding and tree lopping was less likely to provide any equivalent long-term remediation, would cause greater physical damage to the sites, and would be more difficult and costly to undertake); (3) that, should the population be destroyed by fire, other colonies were available within a few hundred metres as sources for translocation or possible natural recolonisation; (4) that burning should take place as late as possible in the summer, allowing the caterpillars to feed for as long as possible preceding their normal low feeding period over winter, and before their food supply was destroyed; (5) burning should take place in the daytime, while the nocturnally active caterpillars were sheltered below ground in ant nests; (6) burning should be as hot as possible to reduce seedbeds of exotic weeds, and to extend to the tree canopy. Further, (7) counts of caterpillars earlier in the season revealed a few 'hotspots' of abundance that should be protected, for example by damping-down. Local community fears were allayed by extensive publicity and the burning was undertaken by experienced local fire brigades as a training exercise with the site perimeters patrolled and monitored thoroughly during the fire. Such exercises can only be undertaken satisfactorily in optimal weather conditions, here windless, hot and dry – to the extent that the burn was postponed for a whole year because such conditions did not prevail over the period initially projected for the operation. Substantial flexibility in organisation is needed, so that any planned burns should not initially be delayed until the latest possible time. Considerations such as the above are needed in each individual case, to evaluate parameters such as intensity, extent, seasonal timing of controlled burns, whether each is contemplated as a one-off or repeated exercise and, if the latter, at what intervals. One practical dilemma is that burning regimes needed to promote host plants may risk direct damage to the insect, and some judicious balance may be needed to achieve the optimal outcome, as above.

It is necessary to distinguish cases of using fire as a management tool for insects for which refuges are available (such as the above caterpillars being able to 'escape' underground) from those in which the insects are most likely to be destroyed by the operation. Thus, prescribed burning of prairie remnants in North America has been implicated strongly in declines of a striking nymphalid, the regal fritillary (*Speyeria idalia*), by eliminating caterpillars (Powell *et al.* 2007). Both contexts require very careful planning, including contingency planning in case of accident, but the latter raises a number of pertinent considerations for conservation: (1) whether refuges of some sort might be provided, for example by mosaic burning and leaving much of the site undisturbed, (2) realistically examining all the opportunities for natural recolonisation of the site from nearby populations, or whether translocations (p. 168) may be needed, and (3) assessing possible alternative measures. The first two of these may be allowed, at least in part, by small scale 'micromosaic burning' (D. Sands, pers. comm. 2008) to provide a patchwork of burns within the larger site. These may be difficult to arrange if a hot burn is needed. Burning areas of only 1–2 hectares, rather than much larger areas, may optimise potential for recolonisation by relatively sedentary insects. Many burning operations undertaken for other purposes (such as fuel reduction in forests) are usually undertaken on a much larger scale, and may not be suitable for conservation of highly localised insects.

On larger areas, far greater flexibility may be possible than on very small sites, simply because greater variation and patchiness in seral succesional stages is likely, but there are few published detailed prescriptions for insect species management that incorporate habitat burning. One major exception is for the Dakota skipper (*Hesperia dacotae*) (p. 123), for which management guidelines include burning, grazing and mowing (Cochrane & Delphey 2002), and for which prescribed burning has also been implicated in loss of populations (COSEWIC 2003). The very specific 'prescribed burn' conditions given as guidelines for managing this species by Cochrane and Delphey (2002) are summarised in Table 6.1, to exemplify the fine level of detail that may need to be defined for any species for which burning is a possible tool for site management. Likewise, prescriptions for haying and grazing regimes for the butterfly were precise – for example that a 20 cm stubble should remain as overwintering protection for the caterpillars, and that floral (nectar) resources should remain undisturbed, so that neither practice is desirable until after the adult flight season. All components should be undertaken on a mosaic

Table 6.1 *Summary of guidelines for prescribed burning as a management component for habitats of the Dakota skipper butterfly*

1. Divide habitat into as many burn units as feasible at the site.
2. Never attempt to burn the entire habitat in any single year.
3. Allow at least 3–4 years before re-burning to allow populations to build up between burns, and generally use the maximum interval adequate to maintain high quality habitat on each unit.
4. Allow fires to burn in a patchy ('fingering') pattern within units, i.e. do not make a concerted effort to burn 'every square inch', unless there is a clear management need for this.
5. Consider use of proactive techniques to increase patchiness of fires, especially when the individual units may be (1) small, (2) greater than 0.5 ha for the burn area or (3) difficult to protect with standard burn techniques.
6. Conduct pre-burn surveys and evaluate other information to indicate distribution and relative abundance of skippers within and among burn units. Because of possible poor conditions prudent to plan surveys for two consecutive years before a planned burn.
7. Spring burns should be as early as possible to limit larval mortality. In contrast, late spring burns may delay/deplete nectar resources for adults, and autumn burns may result in greater larval mortality and exposure of caterpillars to extreme temperatures during winter, through removal of plant material shelters.
8. If fires have to be in late spring (e.g. for control of *Bromus*) other precautionary measures are needed, such as finer scale mosaic burning and allowing for maximum inter-fire interval.
9. If site managed with prescribed fire, subdivide into rotational burn units, not least to help conservation of other butterfly species present.
10. Do not use prescribed fire if smallest feasible burn unit would burn most or all of habitat in one year, unless nearby re-colonisation site is identified. Augmentation/restoration of such adjacent habitat or alternative management (light grazing or late summer haying) may be needed on such small areas.
11. If change is needed in configuration of burn units after prescribing plan review location and timing of recent burns to understand potential effects on current skipper abundance and distribution.
12. Consider any other rare prairie-dependent species on the sites.
13. Plan for 'worst case scenario' of out-of-control fires, and how these might affect the skipper population.
14. Consider reducing fuel levels the previous autumn before conducting burns where fuel levels seem to be high, as high fuel levels increase fire intensity and potential mortality of skippers.

Source: after USFWS (2005).

basis. These regimes are detailed, but also attainable and measurable, in that markers and standards are nominated for guidance.

Although mosaic burning is cited frequently as a management tool for butterflies, predominantly those in prairie or grassland landscapes, Swengel and Swengel (2007) emphasised the values of 'permanent refuges' in areas subject to this treatment, and this consideration may be relevant in any species management programme in which the consequences of burning are highly uncertain – in practice, this means almost all of them! In their broad survey of prairie and savannah sites in Wisconsin (USA) permanent non-fire refugia (with non-intensive managements such as brush-cutting and mowing) within larger areas managed by burning had numerous benefits for a range of ecologically specialised Lepidoptera, and no obviously negative outcomes. Swengel and Swengel pointed out that large numbers of species may be lost when a site is burned for the first time in recent history, with the consequence that 'never-burned' refugia are likely to be even more important for insect conservation than refugia in formerly fire-managed vegetation. Previously burned areas in their study only began to function as refugia 6–8 years after burning. A number of 'listed species' (under ESA, p. 29) had benefited clearly from changes from fire management to mechanical management – and in some cases of the latter, revenue (such as from hay or grazing rights) may even become available to support conservation.

In another useful example of using fire as a management tool for a lycaenid butterfly, Schultz and Crone (1998) incorporated population modelling of Fender's blue in the Oregon prairies, where its habitats can become degraded by weed invasion. The diversity of weeds involved ensures that a strategy to control any one of them may not be effective more generally and, indeed, may facilitate increase of other undesirable plants. The study focused on the control of poison oak (*Rubus discolor*) by fire on the largest known population site, projecting a number of different strategies (Fig. 6.1). Recommended procedure was for burning, on average, one third of the area every year (if funds permit) or every two years (if funds are limiting), a process that yielded the highest long-term population growth rate. Additional long-term field experiments were considered necessary to aid in further refining management.

The details of any such broad management tool such as 'fire' or 'grazing' to maintain or restore favourable habitat for an insect are almost always uncertain, and application may involve a substantial element of trial and error, because of the many variables involved. Wherever this uncertainty exists, treatment should initially not incorporate the whole

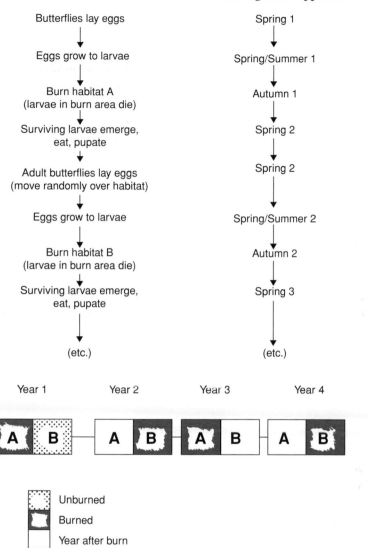

Butterflies lay eggs	Spring 1
Eggs grow to larvae	Spring/Summer 1
Burn habitat A (larvae in burn area die)	Autumn 1
Surviving larvae emerge, eat, pupate	Spring 2
Adult butterflies lay eggs (move randomly over habitat)	Spring 2
Eggs grow to larvae	Spring/Summer 2
Burn habitat B (larvae in burn area die)	Autumn 2
Surviving larvae emerge, eat, pupate	Spring 3
(etc.)	(etc.)

Year 1   Year 2   Year 3   Year 4

A B   A B   A B   A B

Unburned
Burned
Year after burn

Fig. 6.1. Example of assessing management options for a threatened lycaenid butterfly: burning management for Fender's blue (*Icaricia icarioides fenderi*) in North America. Half the habitat is burned each year (bottom), based on the timing of development of the butterfly summarised above this (after Schultz & Crone 1998).

site to be managed, simply as insurance against a catastrophic outcome. Long-term studies may be needed to clarify the precise or supposed effect of a management tactic. Thus, a 16 year surveillance of the rosy marsh moth (*Coenophila subrosea*, Noctuidae), known from a few small raised bog sites in Wales, revealed that early suggestions that maintaining the larval food plant (*Myrica gale*, bog myrtle) depended on fire management were perhaps not true. Rather, hydrological effects alone were sufficient to maintain the plant (Fowles *et al.* 2004). Caterpillar density was, for example, high in a site that had not undergone a fire for at least 30 years.

In many cases, we are dealing with species that are dependent on habitats that are to some extent anthropogenic and altered for grazing or low intensity agriculture, and which have long been subject to some form of 'traditional' management that has proved at least reasonably compatible with species living there. For some insects, then, maintenance or restoration of those traditional agricultural practices may be amongst the best options available. The pamphagine grasshopper *Prionotropis hystrix rhodanica* is a protected species in southern France, where it is threatened from habitat loss by agricultural conversion and quarrying. Being flightless, the grasshopper has very low vagility, with daily dispersal distances suggested to be at most a few metres (Foucart & Lecoq 1998). One possible conservation strategy for this species discussed by Foucart and Lecoq was the need to preserve traditional pastoral practices, particularly sheep grazing during winter and spring, on low-growing (10–20 cm high) 'cousson' vegetation on which *Prionotropis* depends. Former use of the area as an extensive sheep farming common ground was probably of fundamental importance in preventing earlier fragmentation of the occupied habitat into small isolated patches, many of them too small to support the grasshopper for long. Traditional grazing is thus a key element in conserving the habitat.

Grazing variables to be considered in insect management include (1) mode: continuous, rotational, or sessional and, if the last, what seasons and duration; (2) species: either singly or in combination, options vary geographically but may commonly include cattle, sheep, horses, rabbits and more rarely other animals such as alpacas or kangaroos, each with their characteristic grazing modes and trampling effects; (3) extent: whether exclosures are needed on some sites to create a grazing mosaic; and (4) intensity, reflected in density of grazing stock present. Mowing is also a management option on some sites, as a partial analogue to grazing. However, for grass–cutting, a small hand–held brushcutter

M. alcon

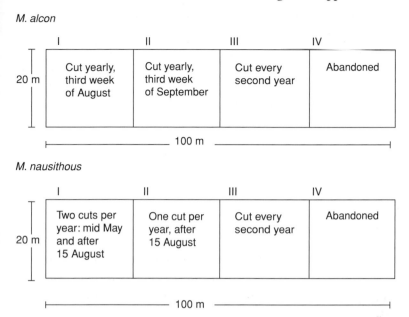

Fig. 6.2. Experimental mowing management for two species of *Maculinea* butterfly (Lycaenidae) in Germany: regimes trialled in contiguous plots (see text, after Grill *et al.* 2008, with kind permission of Springer Science and Business Media).

is sometimes advocated over mowing, if sufficient labour time is available, because it is more versatile and can lead to variable rather than uniform sward heights and a finer mosaic of treatments (Sutton 2006). Likewise, clearing of waterside vegetation to 'open up' areas for dragonflies (but compare with p. 146) is an increasingly common component of riparian management for insects, and one for which the methods also need to be considered carefully. For example, Liley (2005) noted that clearing may normally be done by using chainsaws, but manual use of bowsaws may be needed in wetter areas to prevent water pollution by petrol or chain oils.

As a recent example of experimentally investigating mowing treatment effects, Grill *et al.* (2008) trialled several mowing regimes to promote *Maculinea* butterflies (p. 47) through their influences on the accompanying *Myrmica* ants needed by the caterpillars. Plots in southern Germany were subject to four different mowing treatments based on the seasonal life cycles of different *Maculinea* species (Fig. 6.2). Thus, for *M. alcon* (Fig. 6.2a) the treatments (contiguous in 100 m × 20 m plots) were (1) plots mown in August, (2) plots mown in September, (3) plots mown every second year and (4) abandoned plots. This differed for *M. teleius* and *M. nausithous*, as shown in Fig. 6.2b. Although details of outcome differ for the different ant and butterfly species, the best general management compromise was suspected to be mowing once a year during the second half of September. This was after the caterpillars have left the host plant and entered ant nests, and the treatment enhanced abundance of *Myrmica* spp. in the meadows.

Natural and anthropogenic habitats

Many of the sites managed for insects by burning and/or grazing/mowing noted above are essentially only 'semi-natural' in character because their features are the outcomes of long historical or traditional management by people, and they are now substantially different from their 'original' form. They are now valued widely as the habitats where numerous species of insects live, and on which those insects depend. Many of the key grassland and woodland habitats for butterflies in Britain and elsewhere in Europe fall into this category. In contrast, Australia and some more recently 'Europeanised' parts of the world have a shorter history of substantial change, so that at least part of the insect conservation focus devolves on habitats that can be regarded as relatively pristine (albeit without full knowledge of the changes resulting from earlier human influences). Many of the semi-natural habitats we seek to sustain for insects are plagioclimaxes, and management fundamentally aims to prevent them from undergoing succession toward forests or other climax seres. In essence, conservation depends on maintaining 'disturbance regimes' of various kinds. Long-term maintenance of such ecosystems with a long history of land modification often depends on grazing or disturbances such as fire, so that abandonment of these practices may pose a threat to many insects through facilitating and accelerating habitat change. The impacts of grazing by megaherbivores, particularly on grassland insects, have been studied extensively but, as Ellis (2003) and others have emphasised, detailed knowledge of the effects of grazing on population dynamics of single insect species is often very poor. Such knowledge is commonly based on comparisons of 'grazed' and 'ungrazed' patches at the same site at the same time, or on a single time series from ungrazed to grazed on the same patch. A more detailed study on the bog fritillary butterfly (*Proclossininia eunomia*) over 15 generations in Belgium (Schtickzelle *et al.* 2007) showed that cattle grazing on habitat patches decreased butterfly host plants and led to lower butterfly populations and lower use of the habitat, with markedly increased emigration to other, more suitable, patches within the metapopulation range. However, it was also emphasised that the converse (i.e. no grazing) would fail to conserve the habitat, so that grazing was a pragmatic compromise between this and the impracticable re-establishment of traditional mowing regimes. They noted that grazing regimes should be regulated more effectively – using such aids as mobile electric fences, shelter, and mineral-lick blocks, to prevent both over-grazing and under-grazing. Both seasonal and spatial regulations may be needed.

Habitat re-creation

The extreme case of habitat re-creation may occasionally be necessary, or contemplated, in situations where all available sites are severely degraded and unsuitable for re-introduction or introduction of an insect. Reclamation of derelict land for conservation has long been a strategy in the United Kingdom and parts of Western Europe, but has not yet become widespread in most other places. Re-creation of habitat for particular insect species necessitates very clear definition of requirements and, as Morris *et al.* (1994) emphasised, such ecological engineering necessitates effective collaboration for any specific site, involving planners, engineers, landscape architects, and amenity groups as well as ecologists and entomologists, together with local administrative authority. The aim of any such exercise must be specified very clearly, and be intelligible and agreeable to all parties involved, as must the major actions formulated. Morris *et al.*, in considering the re-creation of early successional communities for butterflies, noted the sequential stages of (1) conceptual reconstruction, (2) detailed planning and (3) implementation. The second of these draws most heavily on ecological knowledge to incorporate adequate sustainable supplies (with sources known and specified) of critical resources and key habitat features within a suitable climatic and topographic arena. As an example of successful mechanical/structural changes to butterfly habitats, creation of 'surface scrapes' for the silver-studded blue (*Plebejus argus*) in an abandoned infilled quarry involved use of a bucket excavator to create 26 scrapes (with total area of around 0.2 ha), ranging in size from 3 m × 16 m to 30 m × 10 m and about 10 cm deep. Topsoil was used to create windbreaks, and this operation led to recolonisation by the host ant (*Lasius alienus*) and larval food plant (*Lotus corniculatus*) (De

Whalley *et al.* 2006). In another brownfield site example, construction of a drystone wall on an old quarry helped to create suitable oviposition substrate for the grizzled skipper (*Pyrgus malvae*) through providing surface for the creeping cinquefoil (*Potentilla repens*) on which eggs are laid (Slater 2007). Clearly, the extent, expense and detail limits the situations in which such mechanically intensive strategy can be considered, and any 'value-adding' (for example to create sites suitable for additional species of concern) as a joint endeavour should be investigated. The example used by Morris *et al.* (1994) considered the requirements of a range of 23 butterfly species on chalk grassland in southern England, but only four of those species were wholly characteristic of that habitat. The exercise might, though, augment specific resources and thereby increase the site's carrying capacity for the other species. The topographical design details for that case are exemplified in Fig. 6.3, and for each feature engineering details were specified – for example, the pattern and method of topsoil removal and re-spreading.

However, the most sobering lesson from this case is that it could not be implemented. Conflict occurred between different sectoral interests, the main opposition being from archaeologists reluctant to allow disturbance of the ground, particularly of the topsoil, and by local conservation/amenity groups concerned over the visual impact of the scheme. The key planning elements devolved on ecological understanding and modest scale, and decisive, practical alteration of land, but failure to proceed emphasised the importance of social and policy aspects of land use and change. Morris *et al.* commented 'Conflict between different sectoral interests in the countryside . . . is inevitable and unavoidable'. In the case of insects, for which much public opinion will not initially be sympathetic, considerable attention to anticipate and reconcile such differences, particularly those involving major change to a site or landscape, is needed in the early planning stages of a project. Social and economic factors may be the ultimate determinants of whether a recovery plan of this extent may proceed, or succeed.

Restoration of any mature natural habitat for an insect is difficult and, as for many other animals, is usually focused on 'framework habitats' to assure supplies of critical resources. For most insects these involve plants, with due regard to requirements of both immature and adult stages of the focal species, or a guild of related species. In local contexts, exercises such as 'butterfly gardening' by individual people can contribute to wider conservation but, in all cases, habitat management is likely to include facets specifically targeting extent and quality of the sites involved. Kirby

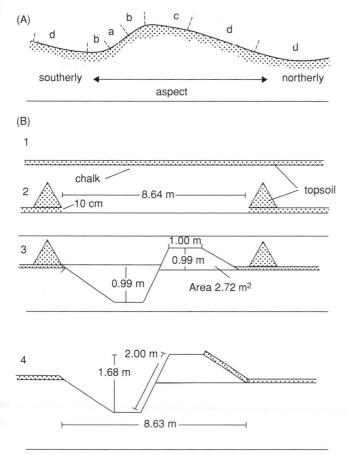

Fig. 6.3. Details of 'ecological engineering' for butterfly habitat: the design and rationale for a project on chalk grassland in southern England by Morris *et al.* (1994), emphasising the importance of local topography. (A) A model of the distribution of microhabitats of the eight rarest butterfly species that could breed on the constructed banks, as (a) very sparse, dry warm turf with much bare ground (*Hesperia comma, Hipparchia semele, Cupido minimus*), (b) sparse warm dry turf (*Lysandra bellargus, L. coridon, Aricia agestis, C. minimus*), (c) short well-drained turf (*L. coridon, A. agestis, C. minimus*), (d) cooler, deeper turf (*Hamearis lucina, Argynnis aglaja*); (B) four stages (1–4) of engineering the banks needed for the above species, with indication of scale, as (1) existing level ground with topsoil overlying chalk, (2) scrape topsoil to sides of working strips; (3) excavate ditch to create banks, (4) spread topsoil back over excavations (Morris *et al.* 1994).

(2001) provided much practical advice for Britain, and much of that transfers easily to other parts of the world.

Some such tactics are advocated widely, and range from general measures likely to benefit numerous species through increasing extent or carrying capacity of habitat to much more specific measures targeting individal specialised species. Simply 'digging ponds' for dragonflies has long been a generally recognised conservation measure for dragonflies (BDS 1988), but very small ponds may not be suitable for particular species. *Leucorrhinia dubia*, for example, may need 'larger ponds' to support a resident population (Beynon & Daguet 2005), and particular gradients or vegetation forms are important variables for other taxa of Odonata. Much relevant background to dragonfly habitat creation was summarised by Corbet (1999), and many species may benefit from such exercises, as well as them having considerable educational and public relations value. This approach is an important general component of conserving dragonfly diversity, reflecting in part that many species are highly dispersive, but 'digging ponds' has been explored only rarely in individual species conservation programmes. Nevertheless, it is a significant option to consider, and the management steps may also have cultural value (for example, in Japan, where some 'ponds for dragonflies' have far-reaching importance: Primack *et al.* 2000). Habitat creation in places beyond the species' current range may become increasingly relevant in the future, should translocations be contemplated to such areas projected to be suitable in the future as climate change eventuates.

Summary

1. Degraded or unsuitable habitats may be rehabilitated for particular insects, as long as the insect's requirements are known adequately as a basis for action. Such 'new' or enhanced habitats are sometimes of critical importance in insect conservation. Restoration may address sites, resources or populations and each is pursued most constructively from a basis of biological understanding of the focal species and its needs.
2. Most habitat restoration for insects involves resources, most commonly plants, either to enhance them as specific foods or to remove them as alien weeds. A variety of management options may need to be considered, in addition to plantings of the former or physical removal of the latter.

3. Procedures such as burning, grazing and mowing are frequent components of habitat management for insects, and considerable experimentation may be needed to determine the optimal regimes for these in relation to scale, frequency, intensity and time of year. Small scale mosaic management may be needed, and considerations also given to competing interests on any particular site.

4. Many key habitats for insects are plagioclimaxes, long managed by traditional practices such as the above. Restoration of traditional agricultural practices may have considerable benefits to insects in such systems.

5. The extreme case of habitat creation, involving ecological engineering to model and construct habitats 'from scratch' is extreme, but may become more important in the future if large scale changes in distributional ranges eventuate.

7 · *Re-introductions and* ex situ *conservation*

Introduction: the need for *ex situ* conservation

Much conservation of insect species takes place on remnant sites where the species still occurs, with the perception that it occurred formerly more widely within the local landscape. In many cases, the target populations are very small (a few hundred or less), and loss of other populations has rendered these increasingly isolated and vulnerable. Under these conditions, on-site ('*in situ*') conservation is usually the first step contemplated, with focus entirely on the resident population and its needs. However, in addition, situations also occur when it is wise to consider one or more of (1) augmenting a field population of the insect from another source, (2) extending the species' range by creating new populations on currently unoccupied sites, and thereby reducing the risk of species extinction, and (3) 'rescuing' individuals or populations either to protect them from high mortality or because the site where they live is to be destroyed. These various needs, and related ones, are outlined and discussed below, as active considerations in many insect management programmes, and for which parallel needs could arise rapidly in others. In some instances, one or other of the above contexts is dictated, but in other instances some choice may be available. Thus, Rout *et al.* (2007) noted that practical steps in a translocation exercise may differ depending on whether the objective is to maximise population size or to maximise the number of surviving populations. Any operation of this kind may be complex and costly but, equally, may be the only or major avenue along which conservation may be pursued.

Re-introduction

Writing on conservation of rare plants, Sutter (1996) commented 'The successful reintroduction of a rare species into a conservation area is a complex and protracted process . . . the process consists of many

Table 7.1 *Outcomes of butterfly releases in Britain and Ireland*

| | Survival (years) | | | | Poorly | Outcome | |
	<3	>3	>10	Reinforcement	documented	awaited	Total
Native species	103	68	21	25	73	30	299
Non-native species	9	1	0	0	12	2	24
Total	112	69	21	25	85	32	323

Source: after Oates and Warren (1990), from Pullin (1996).

overlapping components'. This statement could apply equally to insects. In contrast, Pullin (1996) noted that butterfly restorations (a term largely synonymous with re-introductions) in Britain 'captures the imagination of enthusiasts, probably in part because it appears to represent a quick and simple solution to the problem of species decline', and it is necessary to dispel this as the case in practice. Although the term is sometimes applied rather loosely, 're-introduction' involves introducing insects into sites where the species was formerly known to occur and, thus within the historical range. Related terms are 'introduction', the movement of a species into areas where it has not been known previously or outside its historical range, and 'translocation', a more general term for movement of a species from one place to another, irrespective of whether it was previously known there, or to augment existing populations. Thus, following IUCN (1987), translocation embraces the categories of introduction, re-introduction and 'restocking', the last being building up numbers of individuals in an original habitat, or augmentation.

In the past, many insect re-introductions have been somewhat casual undertakings, many of them without a direct conservation aim, or involving species not of immediate conservation interest. Thus, in their comprehensive review of butterfly releases in Britain, Oates and Warren (1990) (Table 7.1) noted that only 38% of 323 releases were successful (defined by surviving for three years). Many of these cases were poorly documented and the real figure may be as low as 20% (Pullin 1996); extending the criterion of success to Pullin's suggested ten years of persistence, only 12% of releases were successful. Only 47% of those releases had a conservation purpose, for either reinforcement of existing populations or restoration, by founding of new populations. The others included release of surplus breeding stock (29%) and amenity purposes

(17%). Very few of the conservation introductions were documented adequately, and it was historically common practice in such exercises not to monitor releases effectively to determine why they succeeded or failed. Oates and Warren (1990) noted that monitoring was poor or non-existent in about a third of those cases, so that the reasons for their outcome could largely be only inferred, and the kind of information needed to enhance future exercises was not gained. Apparent major reasons for failures included unsuitability of the release sites or lack of appropriate management there, inadequate recording of details so that follow-up actions could not be undertaken, and 'ad hoc' or clandestine releases undertaken without consultation or without permission of the site owners. A major value of Oates and Warren's review is to indicate the core aspects of what should be regarded as good practice, as adopted in several 'codes' from invertebrate conservation bodies – and, conversely, to indicate what not to do. It is valuable, also, in emphasising the need for reporting all attempts: Pavlik (1996, writing on plants) remarked 'No careful attempts at reintroduction are too shallow, no innovations too simple, and no lessons too apparent to go unsummarised and unreported'. So, also, for insects. Apparent initial success may occur. Many butterfly translocations in Europe have failed only after up to 15 generations, with habitat quality the most critical cause of this (Leon-Cortes *et al.* 2003a).

Translocation of insects is an intricate process. As Meads (1994) put it: 'It is seldom possible simply to collect an insect from one place and liberate it in another'. He emphasised the need for systematic planning, with a four-stage process and the skills needed for each stage to be undertaken effectively. Thus, for stage 1 (Fig. 7.1), the skills possibly needed to augment fundamental knowledge at the original site include field survey methods, population estimation methods, and ability to study and evaluate a variety of biological and environmental features. Should captive maintenance (stage 2) be necessary, culture methods and conditions need to be understood, together with observations on behaviour. These needs are common to stage 3, but captive breeding may involve further skills to evaluate longer-term genetic and behavioural changes. The final stage (release) may need to be preceded or accompanied by a range of preparatory techniques and provision for continuing maintenance, together with monitoring.

Nevertheless, direct translocations can sometimes be made, when based on good knowledge of the species and assurance that its needs

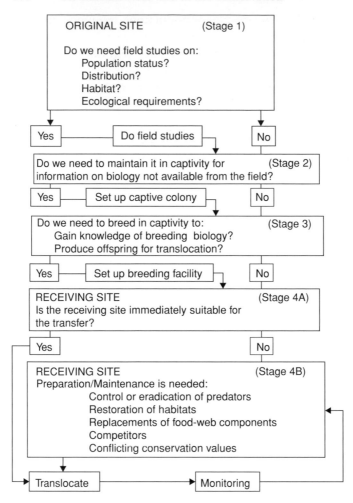

Fig. 7.1. A scheme to facilitate decisions on translocation strategies (for New Zealand weta, after Meads 1994).

are likely to be met at the receiving site. Hochkirch *et al.* (2007) reported one such case involving the field cricket, *Gryllus campestris*, in Germany. A total of 213 wild-caught late instar nymphs were collected from a large population, which had been monitored from 1990 to 2001 (during which time the cricket's abundance had increased approximately 30-fold), stored/transported in boxes containing grass and heather for shelter, and provided with food ('fish food') until release in two sites only a few

Table 7.2 *Factors considered crucial in the success of a translocation project for the field cricket (*Gryllus campestris*) in Germany*

Ecological factors

Habitat quality high in the release area, with suitable management in place.

High habitat heterogeneity in the release area, allowing the insects active choice of suitable areas.

Weather conditions suitable for population growth for several years after the translocation (note: unfavourable weather at first allowed only slow growth; thereafter three 'good years' allowed faster growth).

The cricket is a univoltine species with high fecundity leading to high population growth.

Translocation procedure

Specimens were available from a large wild population rather than from captive-bred stock.

The source population was close to the release areas, so that the individuals were likely to be adapted to local conditions.

Nymphs were used, rather than adults, probably being more efficient.

A high number of individuals was used.

Scientific and administrative factors

Continuous monitoring over a long period of the source population allowed assessment of the security of the donor population, and of the release site confirmed its likely suitability for release.

Excellent cooperation of all parties involved, and at all levels.

Source: after Hochkirch *et al.* (2007).

kilometres away. Subsequent monitoring showed a survival rate of only around 25% in 2002, but a level that the authors considered 'surprisingly high' in view of normal overwintering nymphal mortality. By 2005, the released population and the area occupied had expanded substantially. Hochkirch *et al.* (2007) listed ten factors that they considered of probable importance in this case (Table 7.2), and each of these merits consideration for relevance in any similar translocation exercise. These factors comprised three main categories, reflecting ecological background, the protocol for translocation, and the level of scientific and administrative support for the exercise. Site quality is a paramount consideration for a species translocation but, paradoxically, may for many insects be assessable reasonably confidently only with hindsight, by measuring the success of the introduced population by indices such as the extent and rate of increase.

Box 7.1 · *Insect translocations: what stage(s) to release?*

Very few studies have evaluated the pros and cons of basing releases on different growth stages of insects, but this may clearly have important ramifications and deserves consideration for each species for which trophically different stages and seasonal options are available. For the Karner blue butterfly, Schweitzer (1994) discussed the points for and against releasing caterpillars, pupae and adults, and possible advantages of releasing from the first rather than the second brood. Such inferences may be made for any reasonably well-studied species, for which decisions can be taken on sound biological grounds.

Releases of well-grown larvae were not recommended, partly because of uncertainty of assuring adoption by the *Formica* ant with which they must associate. Rather than risk additional field mortality, Schweitzer recommended continuing to rear the larvae to adulthood. Should it be necessary to release larvae, they should be placed in groups of 50 or more, with one or two caterpillars on each lupin plant, as close as possible to *Formica* nests. If progeny from different females are available, these should be intermingled at each release site, to maximise genetic diversity. Releases of very young larvae may be practicable, preferably on the day of hatching if the weather is fine, again by carefully placing each individual on a lupin leaflet.

Releases of pupae close to emergence (with the adult wing pattern showing) might at times be preferable to releasing adults that might disperse. Pupae should be placed in litter beneath lupin plants, away from direct sun exposure. At least 50 even-aged pupae was recommended as a release unit, and male pupae one or two days older than female pupae may be an advantage. As recommended also for adults (by releasing males a day before females), this would allow males to establish territories before mating. Territory establishment occurs normally within the first day or two of adult life.

Recommended release of adults was for chilled or newly-emerged butterflies on warm (at least 20 °C), dry, and preferably sunny days. Females should not be held for more than 48 h after emergence, as they normally mate while young: a soft release may facilitate mating if numbers are low.

Releases should preferably be in the first brood, partly because progeny survival is likely to be higher than for the second brood, and at the normal flight time for that brood. Augmentation may be possible at the later time of the second brood.

In the future, most insect species for which re-introductions or other releases will be contemplated are likely both to be rare and to have specialised ecological needs. Undue risks cannot be taken with casual or uninformed exercises on such taxa, and species restoration must be linked strongly with ecological understanding and habitat restoration. Suggestions for a butterfly species restoration strategy by Butterfly Conservation (1995) cover many relevant issues (Table 7.3). They include the need for knowledge of site wellbeing and lack of obvious threats to introduced insects, integration of the activities within a broader conservation plan, effective monitoring and communication, and the security of any donor population from which individuals are removed for transfer elsewhere. Removing individuals should not render the donor population more vulnerable. The broader position statement by IUCN (1987) and allied guidelines for re-introductions (IUCN 1995) provide most relevant background information, although not specifically directed at insects.

Several of the cases noted earlier for restoration of insects and the specific plant resources they need demonstrate conspicuous uncertainty over chances of success. Improved restoration strategies are likely to lead to increased success, and also improved chances of detecting and understanding this through monitoring. Although each case needs individual appraisal, published information may provide useful leads. Even the concept of 'success' of restoration poses problems (Pavlik 1996), because of features such as complexity, unpredictability resulting from numerous ecological variables and, in many instances, no clear target or objective endpoint. Short-term goals usually imply that success is positive if a new population is established under conditions in which it can pursue its normal life-history processes in the new environment, and chances of extinction of any augmented population are lowered. For longer-term success, the restored population should integrate with the ecosystem in its new environment, and respond and adapt to environmental changes. Pavlik's (1996) schemata for plants (Figs. 7.2, 7.3) transfer easily to insects but, also, the importance of plant restoration in insect restoration programmes renders their original context valuable. For many insects, plant translocation or other propagation may be integrated with insects being moved within the landscape. The stages shown indicate the parameters that may be considered for monitoring as objectives. We are concerned most immediately with the 'biological success' components of Fig. 7.3, which offer a number of measurable features to fulfil goals and objectives. A further importance of monitoring is to ensure that new biological

Table 7.3 *Main points for a species restoration strategy for butterflies in the United Kingdom*

1. The species should have declined seriously (or be threatened with extinction) at a national or regional level.
2. Remaining natural populations should be conserved effectively, and the restoration plan should be an integral part of a Species Action Plan.
3. The habitat requirements of the species and the reasons for its decline should be broadly known and the cause of extinction on the receptor site (where re-introduction is contemplated) should have been removed. There should be a long-term management plan which will maintain suitable habitat, and the site should be large enough to support a viable population in the medium to long term.
4. Extinction should have been confirmed at the receptor site (at least 5 years recorded absence), the mobility of the target species should be assessed and natural re-establishment should be shown to be unlikely over the next 10–20 years.
5. Opportunities to restore networks of populations or metapopulations are preferable to single site re-introductions (unless the latter is a necessary prelude to the former).
6. Sufficient numbers of individuals should be used in the re-introduction to ensure a reasonable chance of establishing a genetically diverse population.
7. As far as possible the donor stock should be the closest relatives of the original population, and genetic studies should be carried out where doubt exists.
8. The receptor site should be within the recorded historical range of the species.
9. Removal of livestock should not harm the donor population (donor populations may have to be monitored during the re-introduction programme).
10. The re-introduction should not adversely affect other species on the site.
11. If captive bred livestock is used, it should be healthy and genetically diverse (e.g. not normally captive bred for more than two generations).
12. Re-introduced populations should be monitored for at least five years, and contingency plans should be made in case the re-introduction fails, the donor population is adversely affected, or other species are adversely affected.
13. Approval should be obtained from the Conservation Committee of Butterfly Conservation[a] and all other relevant conservation bodies and organisations (including statutory bodies in the case of scheduled species, SSSIs, etc.).
14. Approval must be obtained from the owners of both receptor and donor sites.
15. The entire process should be fully documented and standard record forms completed for Butterfly Conservation[a] and JCCBI[b].

[a] Substitute any equivalent central body or society elsewhere in the world.
[b] Joint Committee for the Conservation of British Insects, now Invertebrate Link.
Source: Butterfly Conservation (1995).

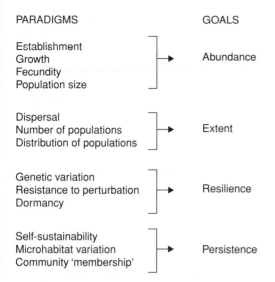

PARADIGMS

GOALS

Establishment
Growth
Fecundity
Population size

Abundance

Dispersal
Number of populations
Distribution of populations

Extent

Genetic variation
Resistance to perturbation
Dormancy

Resilience

Self-sustainability
Microhabitat variation
Community 'membership'

Persistence

Fig. 7.2. The goals of a re-introduction project, drawing on some of the paradigms of success accepted in conservation biology (for rare plants, from Pavlik 1996. This and the following figure adapted from Falk *et al.* (1996), reproduced by permission of Island Press, Washington, D.C.).

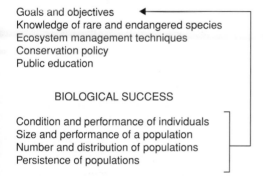

PROJECT SUCCESS

Goals and objectives
Knowledge of rare and endangered species
Ecosystem management techniques
Conservation policy
Public education

BIOLOGICAL SUCCESS

Condition and performance of individuals
Size and performance of a population
Number and distribution of populations
Persistence of populations

Fig. 7.3. The parameters of success of a re-introduction exercise (Pavlik 1996).

knowledge and novel management techniques are documented for future use, irrespective of their success in the current operation.

Captive rearing

Captive breeding of species of conservation interest, for release of reared individuals, is regarded as a highly interventionist form of conservation, with substantial costs involved. It can not be undertaken lightly but, nevertheless, it can on occasion play a pivotal role in management. For many of the most vulnerable insects targeted for conservation, no single wild population may be sufficiently large or secure to permit direct transfer of individuals as foundation for any additional population(s) without incurring additional risk. If any such risk is anticipated, and alternative donor sources are not available, an important ensuing decision is whether establishment of a captive breeding colony, perhaps founded from very few individuals, might be a worthy alternative path, and if so, whether sufficient information exists to undertake this with reasonable confidence. In some cases, also, it might be possible to re-release the parent individuals (e.g. of some butterflies) into the wild after obtaining a complement of their eggs in captivity, rather than remove them permanently. In still other circumstances, imminent or unavoidable loss of key habitat patches may dictate that capturing the stock, as a 'salvage operation', for direct transfer or captive rearing is the *only* possible avenue to conserving a population that otherwise might be doomed. A long-term plan is necessary, not least because it may be necessary to 'hold' a population in captivity for a decade or more, such as when awaiting progress of site restoration to a state suitable for release.

Box 7.2 · *Captive breeding of insects for conservation: some basic considerations*

Ex situ conservation is becoming a recognised strategy in insect conservation, and exercises of this sort are increasing in number, with the ultimate purpose of re-introducing individuals to the wild. However, Snyder *et al.* (1996, writing predominantly on vertebrate programmes) commented that 'Captive breeding should be viewed as a last resort in species recovery and not a prophylactic or long-term solution because of the inexorable genetic and phenotypic changes that occur in captive environments', but also that 'Captive breeding can play a crucial role in recovery of some species for which effective alternatives are unavailable in the short term'. They listed problems as including (1) establishing self-sufficient captive populations; (2) poor success in re-introductions; (3) high costs; (4) pre-emption of other recovery techniques; (5) disease outbreaks; and (6) maintaining administrative continuity over long periods; their additional consideration of 'domestication' is of generally minor significance for insects.

About six main phases occur in conservation breeding for a threatened insect, with the primary aim being to build up numbers of healthy individuals as rapidly and efficiently as possible, in a secure environment. For any exercise anticipated to last for more than about three generations, genetic issues such as inbreeding effects may become important and, where possible, should be avoided by cross-breeding.

The phases are:

1. Knowledge that the wild population has declined, or the sites on which it occurs have changed or are destined for change, to the extent that captive breeding is a desirable (or, even, the only) component of conservation management.
2. Founding a captive population, involving safe capture, transport, housing and maintenance of the foundation stock. Considerable preparation and assurance of food supply may be needed before seeking to establish the population.
3. Growing that population as rapidly as possible, with due attention to increased and often precise needs for housing conditions, temperature, lighting and humidity regimes, food, sanitation and overall costs.
4. Maintaining the captive stock without reduction in quality of individuals over some indeterminate period, commonly of many generations, and instituting monitoring for such 'quality control'.
5. Making releases to the field, with attendant decisions of numbers of individuals, sex ratio, and when and where to release.
6. Managing the released populations and monitoring their establishment and progress. This may determine whether to maintain captive stock over an extended period for reinforcement or repeated release exercises.

For any of these contexts, the principle is that it may be feasible to build up numbers of individuals in protected conditions in captivity, where they are sheltered from the various mortality effects (such as the depredations of predators and parasitoids on immature stages) occurring in the wild. Reared stock, perhaps accumulated over a series of generations until field sites are available and secure, can then be released. Operations of captive breeding (or 'conservation breeding') in this way are still relatively novel for insects, with much of each exercise necessarily experimental and risky, through imposing various forms of 'stress' (including genetic changes through inbreeding; behavioural changes over long periods; changes in individual fitness; conditions for diseases to appear). Most of these conditions will be unknown for any particular insect species, except in very general terms. Wherever possible, it is wise to plan to help to avoid these and related effects by having only short periods (1–3 generations) in captivity, rather than a longer or indefinite period. Changes may occur within a single generation, however (see Crone *et al.* 2007, for discussion), and some parameter(s) of insect 'quality' merit consideration for monitoring in any continuing exercise as an aid to improving husbandry conditions. As one important example, diseases that are largely latent in wild insect populations can emerge in captivity, with a range of pathogenic viruses, protozoans and fungi amongst the most frequent causative agents. Pearce-Kelly *et al.* (1998) recommended that release of endangered insects to the wild should not occur if diseases are detected in captivity, because of risk of transferring disease to the wild population.

Many insects for which captive breeding 'suddenly' becomes an issue in conservation management may never have been reared in such circumstances, so that the exercise is pioneering in both detail and scope. For some insects, advice may be available from local entomologists – many butterfly collectors, for example, have extensive experience with rearing rare species, but their procedures may not have been published. Likewise, practical advice may be available from a local zoo or butterfly house with invertebrate interests. As one common context, mating of the insects may be difficult to achieve in captivity because it may be the culmination of complex behavioural interaction between the individuals in the wild, and that range of natural behaviour cannot be displayed in small cages. One problem with captive rearing of large birdwing butterflies is simply the cost of the large flight cages needed for normal mating to proceed. For some butterflies, it may be possible to 'hand-pair' using methods developed by collectors to rear 'difficult' species or to attempt hybridisations. Clarke and Sheppard (1956) outlined the approach in reporting more than a thousand successful matings with Papilionidae, and noting its success also for selected Pieridae, Nymphalidae and Hesperiidae. For most insects, this approach does not exist.

The allied practice of 'ranching' has been applied for some butterflies, and may also be useful for wider applications. Pioneered for rare birdwing butterflies (Papilionidae) in New Guinea (Parsons 1992, 1998), this technique involves localised high number or high density planting of larval food plants, particularly near places where adults are attracted to oviposit. Reproduction is then concentrated locally rather than being dispersed much more thinly in the wider environment. A proportion of the offspring can then be harvested easily and protected in captivity from natural enemies and other harm. Alternatively, all those found can be reared, and a proportion released into their natal area to sustain the population.

The main circumstances in which a captive breeding programme for an insect may be contemplated include loss of major sites, and the threats to wild populations from overcollecting, diseases and predators. The wider benefits include the knowledge to be gained during the exercise, during which data on fecundity, developmental rates, sex ratio, feeding biology and numerous other aspects can be accumulated and applied also to field management. Indeed, they are a major conduit to the information needed for population viability analyses.

Much of the relevant background to captive breeding of insects has evolved from one or other of three broad sources, namely:

(1) hobbyists' small scale rearing of 'popular' species for personal interest, perhaps particularly of Lepidoptera and some other phytophagous insect groups;
(2) larger-scale rearing of insects for research or pest management, the latter including numerous species of predators and parasitoid wasps for use as biological control agents, and their prey or host species; and
(3) rearing for public exhibition and education, as in zoo exhibits and butterfly houses.

The last of these can be combined with strong public advocacy for conservation. As Pearce-Kelly et al. (2007) noted, relatively modest accommodation requirements for many insects combined with their often substantial fecundity and rapid development allow large numbers to be maintained for modest costs, and material may be made available for field releases within a relatively short time. Problems of disease control and genetic management are, of course, universal concerns inherent in any such operation. The substantial literature on any of the above categories of captive breeding and related husbandry techniques provides much general and specific advice on the practicalities of capture and transport of insects, caging, sanitation, colony maintenance, use of artificial and semi–artificial diets, and environmental control and tolerances. Captive rearing programmes may need to be planned carefully in relation to the duration (cost) and logistic complexity involved. For example, predatory insects can be expensive to rear in large numbers because they may need low density housing to avoid cannibalism, and supplies of living prey which may need to be reared independently. For dragonflies, one possible avenue to eliminating much of this intensive effort is to obtain eggs from captured females and rear the resultant larvae only to some intermediate instar for release, having protected them against the initial high juvenile mortality likely to be a major cause of losses in wild populations.

Box 7.3 · *Insect husbandry for conservation: individual details are important*

Despite the wealth of general information available about keeping and breeding insects in captivity, any ecologically specialised species is likely to present idiosyncrasies that need individual attention without which the operation may be compromised. Various practical themes and questions should be addressed routinely in planning for any such programme, to ensure that adequate facilities and budget are available. Almost inevitably, some risk is incurred as the necessary background information is accumulated so that, wherever possible, a small-scale trial should precede the main programme. In addition to considering short-term maintenance problems, any long-term breeding programme may need also to consider genetic aspects such as likelihood of inbreeding depression, and plan to counter these as well as possible as an important aspect of quality control.

Topics to consider include:

1. Housing; size, number and design of containers needed to allow at least reasonably natural behaviour. Density of individuals may be important in affecting behaviour with crowding, and immature and adult stages may need different conditions.
2. Environmental conditions; predominantly control of temperatures, humidity and photoperiod, with possible need to vary these seasonally or cyclically. Other conditions may occur, for example, aeration of water for aquatic stages or species.
3. Security: quarantine conditions may be needed (or strongly advisable) to prevent escape, or ingress of predators and parasitoids.
4. Sanitation: cleaning regimes and methods, avoiding harmful chemicals and undue disturbance. Frequent cleaning may be needed to prevent appearance of 'latent viruses' or other disease, and vigilance maintained to detect any insect health problems early.

5. Food supply: can this be assured for all insect life stages, all times of the year, and throughout the programme? It may be necessary to grow healthy nursery plants, or maintain stocks of prey organisms, or to include a semi-artificial diet in the feeding regime. Provision of food can become complex and is often more costly than anticipated originally.

6. Maintenance. Adequate curatorial care and responsibility for the wellbeing of the insect population should be assured for the duration of the programme. As an adjunct, good records should be maintained, and will almost always provide original biological information of wider value in conservation and in future rearing exercises. If part of a wider conservation programme, curatorial membership of the management group or continued effective liaison with related interests is necessary.

Strong focus on captive breeding for some taxa has led to this approach (*ex situ* conservation) becoming a major conservation platform. It is, for example, a specified objective for weta in New Zealand (Sherley 1998), where it is noted (p. 13) that 'keeping weta in captivity is technically relatively easy, however, breeding successive generations requires specialized skills and knowledge' with need for research on husbandry techniques for each species involved. Even small practical details of husbandry should be recorded, because this information is an invaluable 'starting point' for considering needs of related species. Thus, Honan's (2007a) manual for the Lord Howe Island stick insect includes notes drawn from experience with the related thorny stick insect (*Eurycantha calcarata*). Importantly, Honan specified details such as features of suitable housing and rearing containers for *D. australis*, and microclimate needs and how to achieve these. He recorded also the results of veterinary examinations of sick and dead insects, and a suite of symptoms attributed to inbreeding depression (Table 7.4), all of which disappeared once additional males were brought in to the colony; hatching levels of eggs, for example, then rose from under 20% to about 80%. Together with the extensive notes given on behaviour, dietary details and reproduction, this manual is an excellent example for emulation.

Examples of rearing programmes discussed by Pearce-Kelly *et al.* (2007) demonstrate approaches for a range of different insect species. A few species, mainly of Lepidoptera, Coleoptera and Orthoptera, have received sufficient attention to lead to production of comprehensive

Table 7.4 *Symptoms attributed to inbreeding depression in captive breeding of the Lord Howe Island stick insect*

1. Unusual morphological abnormalities, particularly on the abdomen of adults.
2. Small egg size and volume, becoming more pronounced with each generation.
3. Low egg hatching rate.
4. Small size of nymphs at hatching.
5. Low survival rate of nymphs.

Source: Honan (2007a).

'propagation manuals' (such as for the Karner blue butterfly, produced by the Toledo Zoo (2002) or even the establishment of units such as the Center for Conservation of the American Burying beetle (*Nicrophorus americanus*) at the St Louis Zoo (2004)).

However, rearing programmes for conservation of many groups of insects have yet to be pioneered, and most of the exercises to date have involved representatives of a rather small suite of orders, mainly Lepidoptera, Coleoptera and Orthoptera and their allies, as above, and some as a specific consequence of actions required by management plans. For example, the conservation plan for the bush cricket known in Britain as the 'wart-biter' (*Decticus verrucivorus*) had a major objective of establishing additional colonies in the wild. A breeding programme at the London Zoo was initiated by using 500 eggs from wild-captured female bush crickets. It yielded more than 3000 eggs after the first year. Problems of cannibalism entailed extra rearing costs, because young nymphs could be kept only in low density groups. New populations were established by release of late-instar nymphs, with successful establishment reported (Shaughnessy & Cheesman 2005). For contrast, a similar programme for the British field cricket (*Gryllus campestris*), which had been reduced to a single small colony in the United Kingdom, has provided more than 17 000 late-instar nymphs for field releases (Pearce-Kelly *et al.* 2007). A captive breeding programme for the Karner blue butterfly (p. 58) (1998–2002) provided nearly 1700 adults for release at a re-introduction site in Ohio, and the butterfly has since expanded in range to occupy the 200 ha reserve (Pearce-Kelly *et al.* 2007).

Some other important programmes have not yet progressed to this stage. That for the Lord Howe Island stick insect (*Dryococelus australis*, p. 98) at the Melbourne Zoo involves a highly specialised monophagous herbivore with a foundation stock of a single pair captured in 2003. This had generated more than 500 individuals by 2007 (Honan 2007b), but field releases will not be contemplated until rats (implicated, as predators, in the earlier decline of the stick insect) have been eliminated from the potential release sites.

Releases

Releases of insects reared in captivity may be planned as a 'one-off' exercise or one to be repeated, for example at generational or annual intervals or over a more extended period with a reserve stock maintained for the duration of the exercise. Even for the first option, it is prudent to retain some component of the parental stock as insurance against otherwise expensive disaster. However well-intentioned the project, and however well the release environment is documented and believed to be 'safe', unexpected adversity is not infrequent. Following from this initial decision of strategy, other decisions are then needed, for example on how many individuals to release, and which growth stages and numbers, what kind of release (hard or soft, as below), at one or more sites, under what weather conditions and at what time of day or year. Each of these topics must be informed by knowledge of the species' biology and by characteristics of the release site(s).

If only small numbers are available, in particular for a re-introduction, a soft release may be preferable. This involves liberation of the insects into enclosures such as field cages erected on the release site. Confinement allows close monitoring of their fate by keeping the individuals in a small circumscribed area rather than allowing them to disperse widely and 'disappear' into the wider environment, possibly without mating because of over-dispersion. In contrast, a 'hard release' involves direct release of insects to the open, without any confinement or control over their movements or fate.

The soft release option may be less important in cases of augmentation, in which the site suitability is already assured through presence of a resident population. It also allows for modifications if the enclosed environment is found to be deficient. Once the caged population is deemed secure, possibly after several generations, the enclosure is removed and the insects allowed to disperse naturally. For species being liberated for the first time, the strategy of utilising only one site and of freeing as many individuals as possible (but without compromising the wellbeing of the captive colony) may be the lowest risk option, rather than spreading the insects more thinly across several sites.

The soft release process may also be wise for direct translocations of field-caught insects to other sites.

Perhaps the world's longest programme on an insect re-introduction has been for the large copper butterfly, *Lycaena dispar*, in Britain, where it became extinct in the mid-nineteenth century. The history of this case was summarised by Pullin *et al.* (1995). Serious re-establishment attempts, involving continental subspecies of the butterfly were initiated in 1926 and the first release (of 38 butterflies) undertaken at Wood-walton Fen in 1927. However, from 1927 to 1955, survival of the butterfly apparently depended on protecting caterpillars from predators and parasitoids by caging them in spring. The population became extinct in 1969 (Duffey & Mason 1970) but release of more than 1000 reared adults in 1970 attempted to re-establish *L. dispar*. Annual population declines necessitated constant replenishment from greenhouse-reared stock (Duffey 1977), with that population eventually maintained in captivity for more than 20 years, with implied loss of genetic variability. Pullin *et al.* (1995) noted that the success of this project is still elusive, and was due in large part to failure to appreciate the true ecological needs of the species.

Understanding the reasons for loss or decline of an insect on a site is a vital precursor to any re-introduction or augmentation exercise planned for that site, but such understanding is rarely complete. The decline of the large blue butterfly, *Maculinea arion*, in Britain (from at least 100 000 adults in the mid-1950s, to one colony of about 250 adults in the early 1970s, thence to extinction: J. Thomas 1995a) reflected increasing unsuitability of the sites on which it had occurred. With enhanced ecological understanding of the species' needs, and especially those of the host red ant (*Myrmica sabuleti*), critical resource and habitat features were clarified and have enabled successful re-introduction of the butterfly to Britain, from a Swedish population. Young caterpillars initially feed on the flowerheads of wild thyme (*Thymus polytrichus*), but older larvae abandon the plants and feed on larvae of the host ant within its nest. Site restoration and preparation has involved scrub clearance (on a site-specific basis, to reduce cover where necessary to 10%–20% and subsequently to manage scrub development by coppicing or burning (on a 5–10 year rotation) and grazing (to maintain the short turf at 2–5 cm high during spring and summer, in which the ant may thrive together with the thyme). The long-term grazing pattern needed is indicated in Fig. 7.4 (J. Thomas 1995a); note that without good grazing management, *M. sabuleti* could be replaced by the unsuitable related species

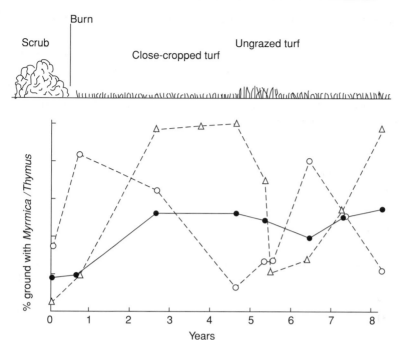

Fig. 7.4. Site management for *Maculinea arion* in Britain: grazing regime effects on the abundance of the suitable host ant (*Myrmica sabuleti*, open triangles), an unsuitable host ant (*M. scabrinodis*, open circles), and the larval food plant (*Thymus*, solid circles) (after J. Thomas (1995a), with kind permission of Springer Science and Business Media).

M. scabrinodis within only a year or so. *M. sabuleti* requires warmer sites, and had itself disappeared from many more sheltered sites to persist only on short turf in south-facing insolated sites in Britain (p. 48). The annual grazing regime needed for *M. arion* sites in Britain must produce short turf in spring and early summer but be removed in mid-June–July to allow the thyme to flower. The kind of animal used for grazing differs with site features. For scrubby sites, ponies or cattle are preferred, whereas calcareous grassland may be grazed by sheep or cattle. Although rabbit grazing may suffice in places, domestic stock are preferred, as they are more easily regulated and the regime is more easily controlled.

Re-introduction of *M. arion* is only one of several similar exercises for species of *Maculinea* in Europe (Wynhoff 1998). Those practical experiences led to enumeration of four recommendations of far wider relevance in insect re-introduction exercises:

1. Before starting a re–introduction, considerable effort should be made to determine whether the characteristics of the release area can guarantee long-term survival of the new population, considering also the possibilities for dispersal and possible need to establish and cater for a metapopulation.
2. Evaluating the re–introduction should go beyond counting or estimating numbers, and also include some 'viability characteristics', such as life span, being determined regularly.
3. Genetic aspects should be integrated into all phases. Thus, it is important to select a large and 'well-functioning' source population, to translocate a large number of individuals where possible, and facilitate good colonisation by immediately increasing the reproductive component of the population.
4. Species-specific management is very important in facilitating colonisation. Once the species has become established and its specific ongoing management established and assured, management may be widened to include other biota on the site(s).

The three major stages are thus preparation (1, above), assurance of quality of the insect (2, 3), and broadening for wider values (4).

Box 7.4 · *Translocating a threatened tiger beetle: practical considerations in a programme for* Cicindela dorsalis dorsalis *in North America*

The Northern Beach tiger beetle (see Box 1.4, p. 36) underwent massive declines over the first half of the twentieth century because of human recreational activities and development of beach habitats in the northeastern United States. Translocations to form new populations are one of the key objectives of the federal recovery plan for the subspecies (USFWS 1993), and the development of a translocation protocol and its trial implementation were discussed by Knisley *et al.* (2005).

Early trials with the beetle involved translocating adults, but did not succeed because these dispersed widely and strongly. Counts every few days showed that the beetles had moved away from the release sites within 1–2 weeks. In efforts to maintain a more cohesive translocated population, Knisley *et al.* used larvae. Initial small scale trials were planned to precede larger operations at a later stage. The beetle life cycle is thought to take two years in the region.

The two beach receptor sites selected had the advantage that they were closed to public access from April to September to protect nesting piping plovers (*Charadrius melodus*). This timing was critical for *C. d. dorsalis*. Larvae are active mainly in late spring and autumn, with adults present in July and August. Contrary to recommendations by Vogler *et al.* (1993, see Box 1.4), the donor populations selected were from the southern part of the range (Chesapeake Bay), because no New England populations were considered large enough to safely donate large numbers of beetle larvae for removal. Beetle larvae (75–150 from each of five sites) were collected by two methods: (1) using a spotlight at night to find larvae at the mouths of their burrows, which were then dug out; and (2) digging from burrows by day. Larvae were transported in individual vials of damp sand, and maintained at low temperatures for 1–2 days, if necessary, until they were released. Releases were made in marked plots, several metres above high tide level, and in rows with larvae a metre apart. The sand was watered (using ocean water) before release, and most larvae began to excavate burrows within only 5–15 minutes. Most larvae were in their third (final) instar, to increase chances of successful pupation and adult emergence in the release sites.

Experimental translocations were made in 1994 and 1995, with subsequent adult counts indicating survival. Larger primary translocations followed in May 1997, 1999 and 2000, leading to peak numbers of adults in July 2001 (749 adults). The beetles appeared to behave normally, and bred. However, numbers decreased substantially through 2002 (142), 2003 (43) and 2004 (6). Reasons for this decline are unknown, but may have included predation by gulls and coastal storm damage.

Knisley *et al.* (2005), based on the initial success of these exercises, recommended that the same methods should be continued, but that closer monitoring at the receptor sites be undertaken to clarify the fate of the population in more detail.

Re-introduction sites

As emphasised above, a major but obvious condition for any insect re-introduction/restoration/translocation operation is that the sites selected for releasing the insects are suitable to support them. Sites for re-introductions/restoration should be within the historical range of the insect (selected on the relatively recent scale of having been known to occur there in the past, rather than longer palaeobiological records!) and, wherever possible, should be those occupied formerly by the insect. Often, this is not the case, and the exercise devolves on other within-range sites deemed similar in character, floristics and environ-ment to those known to support the species. The site(s) must also be secure, ideally protected formally as a reserve within which the focal insect may be among its conservation priorities. Wherever an existing reserve is used, it may be necessary to change previous management aspects to cater for the introduced species in more detail, for example to reduce human access or trampling not previously of any concern.

The wider selection of sites noted above implies needs for physical and climatic compatibility, with flora largely identical to that on previously occupied sites, and adequate populations or supplies of any obligate food plants or mutualistic organisms, or other critical resources. If it is necessary to re-establish or enhance such resources, hydrological and soil conditions for plants must be suitable, and it may be important also to consider availability of any specific mycorrhizas or pollinators needed (Fiedler & Lavern 1996). Some site preparation is a frequent need before introducing sensitive species. Removal of exotic weeds is one such step, but consideration of continuing management (such as by grazing, above) is important at this stage. Overall site value may depend on basic physical factors such as size: some topographically and ecologically suitable sites may simply be too small to support a population without highly intensive management of edge effects, for example.

Re-introduction to historical sites involves careful appraisal of any site changes that have taken place in the intervening period, and which might influence the insect's chances of survival. Thus, the British swallowtail butterfly (*Papilio machaon britannicus*) died out at Wicken Fen in the early 1950s, and changes to the fen had reduced the areas of suitable habitat markedly by that time. In preparation for a re-establishment attempt, large numbers of the larval food plant (*Peucedanum palustre*) were grown from Wicken seed, and about 1500 plants planted there in 1974, as well as around 2000 plants in nearby Adventurer's Fen. Adult butterflies (104 males, 124 females) were released in June 1975, a number considerably lower than had been intended because unseasonal frosts killed many pupae in the outdoor insectary. Dempster and Hall (1980) estimated that about 20 000 eggs were laid over the Fen, and that somewhat more than 2000 caterpillars survived to pupation. Subsequent monitoring, based on direct counts on every plant within permanent transects of 30 m × 1 m, marked an unusually detailed process for such an exercise (p. 201). Declines occurred following an exceptional summer drought in 1976, and the swallowtail had probably become extinct at Wicken by 1979. A major inference for management was that the Fen was basically unsuitable because of the decline in *Peucedanum* and that the major step needed was to make Wicken Fen wetter, by digging peat to lower the land surface to the water table or, less satisfactorily, by pumping water into the fen.

Few studies on insect re-introductions have systematically explored site selection by both resource availability based on detailed ecological information and landscape features to facilitate interconnectedness and support for any metapopulation structure. One such case is for the heath fritillary butterfly (*Mellicta athalia*) in southern England (Holloway *et al.* 2003b), for which sufficient was known about population structure to demonstrate that 'interconnectedness' was a specified condition for increasing the number of populations by re-introductions. These authors produced 'Conservation Strategy Maps' based on GIS information to progressively reveal the most suitable areas, based on habitat, botanical and topographical overlays. Holloway *et al.* considered that this kind of approach to site selection might have much wider application, not least because the many templates can easily be changed to incorporate new information as it arises, or changes in land use patterns. McIntire *et al.* (2007) examined the parameters of a network of sites for restoration to benefit Fender's blue butterfly in Oregon by incorporating the substantial demographic and behavioural information available on this species with landscape maps to simulate connectivity and persistence. Their main conclusion was that restoring all currently degraded and potentially available habitat patches to high quality native prairie would allow long-term persistence of the butterfly, and followed from earlier work failing to adequately predict connectivity between butterfly subpopulations.

Implications of climate change for future changes in species distributions may lead to some change in perspective on 'where to introduce' an insect. Thus, should sites be available at higher latitudes within the established climatic envelope (or historical range) of the species, they may be preferred over those at lower latitudes from which the species may be displaced earlier and become climatically stressed even sooner. Such uncertainties emphasise again the highly experimental character of much insect conservation work on this topic.

Summary

1. Some insect conservation programmes necessitate aspects of re-introducing species to sites where they occurred previously, or augmenting small populations by release of additional individuals. These

can be taken from other field populations (if sufficiently robust populations exist), or be derived from captive bred stock. Such '*ex situ*' conservation is an important component of some insect species conservation exercises.

2. Re-introducing an insect species is a complex process, needing careful planning in each individual case and to assure the quality and suitability of stock available. Success of any insect release or translocation depends on receptor site quality. Monitoring of all insect re-introduction and translocation exercises has potential to add considerably to understanding why an attempt succeeds or fails.

3. Captive rearing of ecologically specialised insects is also a complex undertaking, and preliminary 'trial and error' research may be needed before committing to a large scale exercise involving a threatened species. Particularly when it is anticipated that the species will be bred over many generations in captivity (for example, to await successional recovery following preparation of potential release sites), precautions to counter genetic and other deterioration of quality may be needed.

4. Releases of insects also need careful planning and consultation. Relevant practical aspects to consider include numbers, stages and sex ratio of individuals for release; seasonal and diurnal timing of release; whether the exercise is a single one or to be repeated at intervals (generations, years); to be a hard or soft release; and mode and frequency of post-release monitoring. If releases are made from captive stock, it is prudent to retain the stock until the exercise has been completed.

5. Preparation of receptor sites must draw on the best available knowledge of the species' requirements, particularly to ensure that (1) adequate supplies of critical resources are present and (2) previously known or operative threats are absent.

8 · Roles of monitoring in conservation management

Introduction: the need for monitoring

There is danger that any conservation management plan will remain rigid and non-responsive to changes that occur to the species, population or environment being managed. It may thereby lose the initial perspective and focus as the operating environment diverges from the basis on which the plan was formulated. Changes in the species' conservation status or in its environment may result from management or from other, non-anticipated, factors. Ideally, management should be adaptive and either periodically or continuously dynamic in response to review as such changes occur. It follows that those changes must be detected and interpreted as reliably as possible to enable refinement of management, and curtailment of ineffective management components. As Hauser *et al.* (2006) noted, monitoring programmes are planned most commonly with the assumption that monitoring must be undertaken at fixed time intervals. For insects, these intervals are usually annual, to coincide with accessibility to a conspicuous life stage amenable to detection and counting. Often, these correspond to an intergenerational interval.

Monitoring is therefore sometimes defined as 'intermittent or periodic surveillance', and is the major means through which the success or failure of a conservation management programme can be assessed. It becomes particularly important when information is needed on short term effects of management, because correspondingly short term revision of that management may be critical if adverse effects are found. It also has many wider applications in revealing trends in abundance and distribution, some of which may trigger additional conservation interest or activity. Quantitative observations over time (so-called 'longitudinal studies') are often the only way in which the success or otherwise of management can be assessed. However, comprehensive or long-term monitoring is expensive, and considerable efforts have been made to reduce those costs

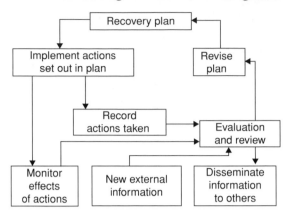

Fig. 8.1. The place of monitoring and how it should influence management plans in conservation (after Ausden 2007, by permission of Oxford University Press).

by focusing monitoring exercises carefully. In a context rather different from the single species studies of primary concern here, use of butterfly assemblages as indicators of forest logging in Borneo, Cleary (2004) noted that savings could be made by monitoring genera (rather than species) as more easily separable than species by sightings alone, so that any accidental damage from captures is avoided. Visual conspicuousness greatly facilitates insect monitoring.

The major purpose of monitoring is to provide information that can be used to formulate, and later refine, conservation management. Initially, it might be needed to help allocate the focal species to a category of threat. There, as in other contexts, the precise monitoring objective may influence the outcome (Joseph *et al.* 2006). As Sutter (1996) emphasised, monitoring involves repeated measures, samples, or inspections to determine changes in the abundance and distribution of a population, species or one or more key environmental attributes over time, to detect natural trends or responses to management. Wherever possible, these should be by ways or indices that are quantitative, and readily measurable. Monitoring has the twin roles of providing initial (baseline) and later biological information and, as a component of management in conservation, is necessary to enable that management to be dynamic. Ideally monitoring of insects should never involve destruction of the insects, so that many forms of trapping insects for inspection are excluded as monitoring tools. Monitoring is a long term exercise, and should be planned as such, and needed both to assess the success (or otherwise) of management, and to detect new problems as they arise (Fig. 8.1). The precise purpose of monitoring must be defined carefully for each individual project, simply to focus on the most useful and economical aspects, and to avoid tangential efforts. In short, monitoring is accepted widely as a key component of conservation management (Lovett et al. 2007), but must be based on very clear questions and be designed carefully in relation to the key variables to be measured, rather than simply lead to accumulation of data with little compelling focus and unknown purpose.

For example, in re-introduction or restoration projects, it is vital to know whether the biological objectives (p. 175) are being met. The initial question may be simply whether the introduction has 'worked', and later be followed up by knowledge of increased numbers and distribution, and other aspects of the species' 'performance' in the new environment. Unsuspected factors commonly arise: generalist predators not earlier considered of importance may reveal themselves to be a significant cause of mortality, for example, and such components can be detected only by inspections. Ideally, monitoring should be a continuing long-term process, with commitment needed at least for the duration of the project. Unfortunately, though, the project itself is often too short to accommodate this need. Funding cycles of 3–5 years drive many projects in conservation, so that there is commonly no guarantee of continuity beyond that span, however compelling the need.

Where long-term monitoring has indeed been possible it has provided some of the most convincing data on insect declines and urgency of conservation need. The major component of such information on status changes in British butterflies is based on recording using a 10 km × 10 km grid across the entire United Kingdom (Asher et al. 2001), and constitutes one of the world's most informative monitoring schemes of an insect group. Used as a temporal baseline, it has revealed many changes in species distributions and abundances over much of the last century or so, and allowed inferences on the causes of those changes. The Rothamsted Insect Survey has coordinated a network series of light traps in Britain since 1968, and the population trends for 338 species of moth over 35 years were analysed recently by Conrad et al. (2004). The large scale of that survey is unusual. As Conrad et al. commented, appreciation of the values of such long-term data sets is increasing, although emphasis remains on short-term projects, not least because of practical difficulties of renewing funding for longer periods. The Rothamsted survey depends largely on participation of volunteers who provide records from their personal light traps. Over that period, the proportion of moth species displaying significant decrease (54%) was more than double that displaying increases (22%). Parallel surveys elsewhere in the world would be a significant investment in helping to set priorities for future conservation strategies for insects.

Many commentators have emphasised the care needed in ecological monitoring, and that undertaking the exercise without rigorous controls on standards can be largely a waste of time. Field *et al.* (2007) suggested three general problems that apply to many monitoring contexts.

1. Funding commitment must extend for a period sufficient to allow a change to be detected and differentiated clearly from normal temporal fluctuations. For insect population numbers, this may represent a considerable period because of commonly large (but almost invariably unknown) natural changes that occur. Field *et al.* suggested that ten years is a 'sensible minimum target' for ecological monitoring programmes. Changes in distribution of an insect may be detected considerably more rapidly but, again, these may simply represent temporary range expansions or contractions resulting from unknown population structure and dynamics.
2. Objectives must be stated clearly, so that both the variable and the extent of change must be specified.
3. The sampling design must have sufficient statistical power to detect a change, if it actually occurs. Field *et al.* (2007) recommended that early results (for example, from the first or first few monitoring occasions or seasons) should be analysed promptly as a guide to whether refinement in approach is needed.

Criteria for monitoring

Flowing from the above, four basic criteria are needed in monitoring (Sutter 1996), not all of which are always easy to accommodate.

1. Data must have a known level of precision, so as to detect change, rather than just 'noise'. In practice, greater precision may mean 'more expensive' because more sampling will be needed, particularly if the insect or resource is dispersed thinly. A major decision is whether presence/absence information is sufficient (for example, so as to determine establishment of a translocation) or whether given levels of numerical change or trend are required to fulfil specific objectives of the plan.
2. The methods used must be standardised and repeatable on every monitoring occasion, as with any other form of ecological sampling. With

long-term projects, it is unlikely that the same personnel will always participate, so exact details of sites and monitoring points (for example through GPS coordinates and photographs), timing and numbers of inspections and the precise methodology employed (Samways *et al.* 2009) must all be recorded. Any maps, together with the accumulated data sets, should be archived responsibly and assured accessible for future work.

3. As implied above, responsible monitoring on many projects is likely to become a long-term exercise, particularly to determine effects of management. Despite the short generation times of many insects, population responses to environmental changes may take several years, at least, to manifest.

4. Overall costs must be maintained as 'reasonable'. Demand for excessive detail or greater numbers of monitoring occasions may not be supportable, even with the welcome help of volunteers where they are available. This reality obliges monitoring to be focused clearly on specific questions and contexts, rather than becoming a generalised survey exercise. For many insects and plants, for example, an annual monitoring exercise may be sufficient. For others it may be useful to monitor more often to help establish initial population or life history data, and several counts each season may then be needed to cover an extended flight period or to inspect different life history stages. Populations of the golden sun-moth (*Synemon plana*, p. 30) can be evaluated only from multiple counts, for example, because any single occasion count can reveal only a small part of the resident population on a site (Gibson & New 2007).

In essence, the above constraints and conditions mean that the objectives of insect monitoring must be specified clearly in the planning stage, as should options for future change in response to what is detected. A 'flow chart' of management possibilities in relation to various monitoring outcomes is both an instructive initial exercise and a valuable practical tool as the programme proceeds (see p. 55).

At the outset, it is necessary to address several basic points:

1. What is the central purpose of the monitoring exercise? Is the proposed methodology robust and suitable for this?

2. What is to be monitored? If the insect, is monitoring to be based on numbers, distribution, or some form of 'fitness' or population feature (such as recruitment)? If the habitat or some resource(s), are amounts, distributions, quality or other (defined) parameters targeted?

3. Why is that/those parameter(s) targeted? It is extraordinarily easy to be lulled into monitoring features that are not necessarily related to clear conservation outcomes but are 'easy' to inspect. It follows that objectives must be kept very clearly in mind, and tangential aspects restricted.

4. Can the proposed monitoring programme be modified, easily and without compromising its primary purpose, to provide additional information of value for understanding management – for example by stratifying the sample points in relation to habitat or wider resource characteristics or gradients?

Approaches to monitoring

A range of approaches to monitoring may be available, but differ according to the primary objective of the exercise. Determining presence/absence of an insect, for example, may be based on a standard period of observation or search, which may be undertaken at fixed intervals, annually or on repeated occasions dictated by availability of the monitoring target. At its simplest, a single visit to a site during a predictable flight period may be all that is needed to confirm the insect's presence. Other cases may be less predictable. For the golden sun-moth (p. 30), a recommended observational protocol is for up to four visits to a site over the flight period of the moth to increase chances of finding the male moths (the only stage amenable to easy detection) under suitable conditions (Gibson & New 2007), with later visits abandoned if the moth is found. If possible (and often this is not the case) a reference site on which any highly seasonal species occurs abundantly in the same general area as sites to be monitored can be used to reduce sampling effort: detecting the species there implies strong likelihood of it being findable elsewhere at around the same time – if it indeed occurs there. Across any wide range, phenological vagaries (p. 185) must be expected, and sufficient sampling latitude incorporated to undertake multiple visits to monitoring sites.

Protocols for such non-quantitative searches should be disciplined, but this is not always as important as when quantitative information is sought. The latter, such as to determine trends in population size, is far more laborious to obtain, particularly as any trend necessitates monitoring intensively over at least several seasons or generations, and the extent of background 'normal fluctuations' themselves are likely to be unknown. Nevertheless, some realistic standardised 'giving-up time' may be adopted, to represent standard sampling effort when searching for a species unsuccessfully. Many monitoring exercises involving flying insects are based on timed point counts or on timed or measured transect walks (see Pollard and Yates (1993) for background). The latter can enable larger areas of habitat to be inspected within a given period, and also allow for some compartmentalisation of habitat quality by extending across different major habitat within a site. The Pollard and Yates technique involves the observer counting insects within 5 m of the transect line, but some applications extend the area to be surveyed considerably. Thus, surveys for the large and conspicuous regal fritillary butterfly (*Speyeria idalia*) by Powell *et al.* (2007) involved visual counts to about 30 m each side of the central line, with statistical corrections to assess variation of detectability with increasing distance away. One such cause of bias involved higher disturbance and counts caused by 'flushing' butterflies from vegetation as the observer walked along. An alternative technique, of point counts, focuses more intensively on selected parts of the habitat rather then the extensive area, and points can be randomly selected or chosen by reference to particular environmental attributes affecting likelihood of occurrence or abundance. As before, a standard observational period or transect distance is needed before declaring a species absent or halting a count.

Although devised for butterflies, transect walks can be applied to a considerable variety of other conspicuous insects, sometimes with minor modifications to accommodate their biology. Ideally such a monitoring method should be (1) sufficiently general that it can be applied and replicated easily in time and space; (2) sufficiently robust but also sufficiently flexible to be used with little change for a range of taxa and habitats; and (3) usable for repeated measures at the same site and for direct comparisons with other sites. Care may be needed when different people participate in visual monitoring of insects, because they may differ considerably in their observational powers and experience; if quantitative data are the primary focus, it may be necessary to formulate some form of trials on inter-observer calibration to determine whether this difference might provide an unwanted bias.

For many less conspicuous insects, more intensive searches may be needed, and focus more on habitat, such as by sifting litter, sweep sampling of low vegetation, or involve use of some form of bait. Light trapping for moth monitoring is common, and many standard or more unusual insect sampling techniques may have applications in monitoring exercises.

Mark–release–recapture (MRR) methods have considerable value in population studies, but can be laborious to undertake and are not applicable easily to all species. If such methods are contemplated, expert advice on procedure, suitability and analysis should be sought. Applications of MRR include measuring population size, detecting dispersal or colonisation, determining individual longevity, change in sex ratio over time, and appraising the duration of a flight season or life stage. However, practical difficulties may include low density or high dispersiveness and hence low chance of recapture, and that inexperienced handling of small delicate insects can easily result in damage or death, so that considerable care is needed. It is a presumption of many MRR methods that marked insects behave normally after release and are no more susceptible to predation and other external influences than are unmarked individuals.

Box 8.1 · *Principles of mark–release–recapture (MRR) for insect population studies*

The basic operation of MRR methods is that individuals are captured from a population, marked in some way that will facilitate their future recognition (either individually or by batch marking on the same occasion), and released unharmed to disperse within the population. On one or more later occasions, insects are captured randomly from the same population and the proportion of marked to unmarked individuals used to calculate population size. At that time, unmarked individuals may be marked in a manner that distinguishes them from those marked earlier, and the process repeated. A series of such observations can provide much information of biological interest: on changes in population size over the period, individual longevity, sex ratio, and (with individual markings) distances moved and any such differences between the sexes. Population size estimates depend on the population being closed, but different markings for insects at different sites can provide evidence of inter-site movements and, perhaps, indicate a metapopulation structure.

Various indices can be used to calculate population size and rate or extent of individual turnover within the population, and these are summarised in texts such as those by Begon (1979) and Southwood and Henderson (2000). However, the success of MRR methods depends on several assumptions being validated, and all of the following must be considered carefully in contemplating an exercise involving this approach.

1. The insects are not harmed by marking, and their subsequent behaviour and vulnerability to predators should not be affected by the capture, handling and marking process.
2. The marks must be sufficiently durable to allow the study to proceed, so (for example) should not be washed off by rain, and interpretable in the future. 'Batch marks' (in which it is not necessary to distinguish each individual insect) may be simply different colours on each occasion; 'individual marks' necessitate either sequential numbers (as can be written on a butterfly wing) or a code of different spots, perhaps of different colours and positions on the insect.
3. The marked insects should be able to disperse freely after release, and release should be (as far as possible) at or near the point of initial capture. It may be necessary to cage marked insects for a short time before release, for example to ensure they have not been injured or had their wings stuck together by wet paint/ink, or if the weather has become unsuitable.
4. The probability of capturing a marked insect must thereafter be the same as that for any other member of the population. In practice this may mean that the operator must consciously resist a temptation to select conspicuous marked individuals.
5. The population is sampled at discrete time intervals and the duration of sampling on each occasion standardised (for example, for time of day and weather conditions).

Monitoring indices for resource abundance and quality must also be defined very carefully, and the aim is then to determine whether particular levels or standards have been attained, or whether the trends or trajectories towards those desirable standards are favourable.

Studies of published monitoring programmes for insects and their requirements are strongly advised as an aid in possibly refining a draft protocol for a given species once the initial objectives have been defined. There are many possible exemplars. Most commonly, a single intergenerational 'marker' is used for insect monitoring – normally this is the adult insect as the most conspicuous, easily recognisable and easily assessable life stage. More intensive or repeated monitoring can become expensive, and any need for this in management monitoring (as opposed to basic scientific study of the species) should be appraised very carefully in terms of the commitment it engenders. Additional monitoring is warranted

Table 8.1 *Monitoring data for Edith's checkerspot butterfly: caterpillar population estimates and distribution by site microclimate*

The range given is the 95% confidence interval based on the normal approximation.

	Habitat area	1985	1986	1987	1988
Total area	89.32 ha				
Total number of caterpillars		92 000 ± 27 500	472 000 ± 56 500	783 000 ± 81 000	319 000 ± 36 000
Very warm	7%	1%	1%	2%	1%
Warm	10%	3%	8%	9%	19%
Moderate	48%	19%	33%	59%	43%
Cool	21%	31%	36%	21%	28%
Very cool	14%	46%	22%	9%	9%

Source: after Murphy and Weiss (1988).

only when it addresses specific focused questions. Thus the adult is often by far the most mobile and dispersive stage, so that adult counts of a butterfly, for example, may give good data on population size, but none on patterns of resource use by caterpillars. Separate counts of caterpillars (as for the Eltham copper: Canzano *et al.* 2007) may reveal patterns of distribution across food plants of differing qualities, for example. In this subspecies, parallel surveys of adults (by transect-based counts) and caterpillar counts (direct searches of all food plants in selected 10 m × 10 m quadrats) are undertaken as complementary measures of population size and micro-distribution.

Late instar caterpillars of the Bay checkerspot butterfly (*Euphydryas editha bayensis*) were assessed by Murphy and Weiss (1988). In this species, these larger caterpillars are conspicuous as they bask and feed among short grassland vegetation, and they were counted in series of 50–100 quadrats 1 m × 1 m across a 100 ha habitat patch that supported a very high butterfly population. Counts were related to topographical details to evaluate differences with aspect and slope and thereby some effects of developmental temperature regimes. Some of their results (Table 8.1) showed disproportional changes across microsites during the four year sampling period. As the population of larvae increased about eightfold, the proportion on 'very cool' slopes decreased from about 46% to 9%. Over the three years from 1985 to 1987 the proportion of larvae increased on 'moderate' slopes (19% to 59%) and on 'warm or very warm' slopes (4% to 11%). Decreased populations in 1988 reflected the host plant growing season being curtailed by lack of rain in 1986–87, so that a clear link with resource supply was evident. The methodology proposed by Murphy and Weiss (1988) included a number of conditions of very wide relevance in insect monitoring. The technique (1) is non–intrusive (insects not handled) and low impact (minimal damage to the habitat), (2) is repeatable, (3) gives absolute rather than relative population estimates, (4) documents demographic processes that are responsible for year-to-year fluctuations in population size, (5) provides a baseline for future monitoring and mapping of topographic features that contribute to habitat quality, and (6) is labour-efficient and, by using simple approaches, is low cost. The entire sampling exercise could be undertaken by one person in only five field days. The approach represented a stratified sampling by microclimate, and this was facilitated by the habitat being low-growing vegetation. Other environmental axes might be distinguished for study in cases where vegetation is more complex – for example, with more vertical layers.

The post-release monitoring of *Papilio machaon* on Wicken Fen (p. 188; Dempster & Hall 1980) also involved systematic searching of plants on defined permanent transects (four transects in 1975 and 1976; increased to five transects in 1978 and 1979 after insect numbers declined). This case was unusual in its high sampling intensity, this being undertaken once or twice a week to enable detection of all stages (eggs and each of the five larval instars) to produce a life table, or mortality schedule. It thereby paralleled the more intensive sampling more common in determining economic threshold levels of agricultural pest insects. However, Dempster and Hall noted that this intensity of study 'opened up pathways within the vegetation' for operator access, so that oviposition data might have been increased by giving searching female swallowtails easy access to low-growing *Peucedanum* that they might not otherwise find. The intensity of sampling in a monitoring exercise can thus become problematical, and must be considered carefully in relation to the value of additional information obtained, or the likelihood of this being in some way 'distorted' or even harmful to the species. The population data obtained may be invaluable in interpreting population dynamics.

In some monitoring exercises, it may not be necessary to actually see the insects, if specific characteristics or traces are available. Thus, Hochkirch *et al.* (2007) monitored their translocated field cricket (p. 170) population growth by counting the number of singing males on each site. The song can be heard up to 100 m away and, because some males will not be singing at any given time, the values obtained will represent minimum number but give an index likely to be valid over a time series when sampling is undertaken in suitable conditions (dry, warm and windless days during the main calling season of May–June). For some insects, neither adult nor immature stages are easily or directly available for inspection, and some form of even more indirect or surrogate measurement may be the only possible option. This may be straightforward: many plant-feeding insects, for example, leave very characteristic feeding signs. As examples, caterpillars of *Hesperilla flavescens* (p. 38) in Australia weave leaves of *Gahnia filum* together to construct retreats (Crosby 1990), and those of the frosted elfin butterfly (*Callophrys irus*) in North America gnaw a 'feeding ring' on wild indigo plants (Albanese *et al.* 2007, 2008). Direct counts of these are valid surrogates for counts of the insects, but in the case of 'retreats' give no estimate of mortality. As with counts of galls, leaf mines, or similar signs from specialised feeding habits, they are a valuable 'first approximation', but further examination may be needed to validate some interpretations and applications. If monitoring normally cryptic insects, such as parasitoids, it may be necessary to seek features such as emergence holes and also to bring sample series of hosts to the laboratory for direct rearing of these insects. The last approach needs careful consideration, with care not to deplete the field population to an unsafe level. Likewise, the extent of persistence of such general features should be assessed. 'Old' galls, with the inhabitants long emerged, may persist for up to several years in some cases, for example, and leaf mines on evergreens can last well beyond one generation time of the causative insect. Uncritical counts in such cases may confound cross-generational information; in others, 'fresh' structures (such as the *Hesperilla* retreats on *Gahnia filum* (p. 38)) are obvious.

In extensive programmes, it may not be possible to monitor all populations affected or of interest, not least because of the site's fragility, so that small habitat patches may suffer from visitations, such as by trampling or other disturbance.

For Hine's emerald dragonfly (p. 71), USFWS (2005) noted that tracking population trends of all populations is desirable, but logistic constraints may prevent this. Several key sites were proposed for intensive annual census surveys, and simple annual inspection for presence on other sites was recommended. The suggested rationale for annual intensive monitoring was to provide information on population 'health' and size trends, and also to try to link larval and adult population size estimates through surveys at one of the 'subpopulations' in each monitored area. USFWS (2005) noted the need to develop suitable methods for this, and the advantages of larval surveys are that these can be undertaken in inclement weather and over a longer period than is available for the adults. Counts of exuviae (the cast skins of last instar larvae left on emergent vegetation when the adults emerge) have been used extensively elsewhere for population studies of Odonata. Exuviae can be valuable as being relatively conspicuous, persistent and with numbers related to features of the adjacent aquatic habitats (see Ott *et al.* 2007, on the European *Oxygastra curtisii*).

In an ideal outcome, analysis of long-term monitoring data can (1) demonstrate the outcome or trajectory of management, with the optimal result being that recovery to anticipated levels has occurred, and (2) help to understand why that outcome has eventuated. The former evaluates the success or otherwise of the programme; the latter is important in helping to understand why the outcome occurred, and can be valuable also in analysing other cases. Although, as noted earlier, it may be dangerous to transfer biological knowledge uncritically to other populations or sites involving the same species – as well as to different species – the background obtained from a single programme can provide valuable clues or 'hints' for wider consideration. A recent summation for the chalkhill blue butterfly (*Polyommatus coridon*) in Britain (Brereton *et al.* 2008) involved annual transect monitoring of butterflies on 161 sites from 1981 to 2000, so overcoming reliance on single site features, and revealed a threefold population recovery over that period. All increases were at established sites, and no re-colonisation or range expansions were detected. This programme was not accompanied by a formal national Species Action Plan, but generic objectives under the UK Biodiversity Action Plan were (1) to halt the decline in the short term and (2) to restore the 1950s range in the longer term. Most of the population increase occurred in the 1980s, and four possible causes of this were cited by Brereton *et al.* The first, range expansion, was discounted as not having been detected. The others were (2) favourable weather, (3) changes in autecology and/or (4) habitat change at established sites. Warm weather in conjunction with grazing management appeared to favour population increase. Although the precise mechanism of this is not known, the effect may be linked with increasing breeding success of adults and shortening developmental times, together with increased supply of the larval food plant (horse shoe vetch, *Hippocrepis comosa*). Habitat condition changes were believed to be the most important influence on population increases, and result from a range of factors: (1) recovery of rabbit numbers, (2) increased stock grazing levels (cf. p. 185), (3) conservation designation and management (with many sites designated as 'protected' in some way, and including an increased number of nature reserves), and (4) participation in agri-environment schemes (as a policy aspect, with management largely preventing butterfly declines in the1990s).

Summary

1. The practical steps taken in any insect conservation plan must be monitored carefully to determine the success or otherwise of management against measurable defined criteria. For many insects, a 'monitorable stage' (such as a conspicuous adult) may be available for only a few weeks in any year, and monitoring must be focused on that, commonly predictable, opportunity.
2. The major monitoring parameters are population size, and distribution of the focal insect, and the supply and quality of resources (or other key environmental variables) with which it is associated.
3. Monitoring must be non-destructive and, as far as possible, long term, but must achieve far more than simply accumulate data. It can do so only if it is designed carefully in relation to specific questions to be addressed, and to the insect's biology.
4. Criteria for monitoring must, likewise, be realistic rather than allowing the process to become 'open-ended', and a variety of techniques and approaches are available to assure adequate standardisation. A sound monitoring protocol for an insect must incorporate considerations of scale sufficient to encompass the species' biology in terms of (1) broad aspects of frequency, timing and standard observations; (2) considerations of time of day, weather conditions and relative acuity of participant observers; and (3) extent of information sought (presence, population numbers, relative abundance across habitat patches, and others).
5. Conservation management must be adaptive, or responsive to the findings of monitoring.

9 · *Insect species as ambassadors for conservation*

Introduction: extending the conservation message

Many insects still have 'novelty value' in conservation, with demonstration of their needs and intricate ecology having potential to stimulate interest (and, even, wonderment) as people are made aware of these. Because of this, every campaign for conservation of an insect species may contribute to wider knowledge and understanding, and help to raise awareness of the relevance of insects in the natural world. Each such species may become an ambassador or flagship for wider conservation need, and acceptance of such roles may become an important part of gaining support for management. Education and fostering wider interests are thus important components of wider conservation management for insects, and opportunities to pursue these should be considered from the initial planning stages. The practical help ensuing, such as garnering enthusiastic volunteers and related support, may prove critical to the success of a conservation plan. Such practical aid is distributed very unevenly, and is available much more readily in temperate regions than over much of the tropics, reflecting the geography of species-level conservation interest for insects noted in Chapter 1.

Temperate region insects

A number of insect species conservation efforts have involved insects that have indeed become notorious causes célèbres, as flagships and potent ambassadors for increasing appreciation of conservation need in various parts of the world. These are almost all in temperate region countries, where the knowledge, capability and goodwill to conserve insects can be well established, as displayed through many of the examples cited earlier. Some of these have been central considerations in conflict over use of particular sites designated for developments. In such cases, the 'political aspects' of conservation tend, at least initially, to predominate over 'biology', because of the urgency of saving the site from despoliation as a basis for all future conservation management for the species. As in other contexts in which public interest is a prime influencing factor, 'popular' insects, predominantly butterflies, are the most common central characters in such conflict. Nevertheless, other insects may become involved from time to time. The Delhi Sands flower-loving fly (*Rhaphiomidas terminatus abdominalis*, Mydidae), the only species of Diptera listed as Endangered in the United States, occurs only on small remnants of a particular sand dune ecosystem in southern California. This land has high economic value, and the fly has been important in preventing development of an industrial enterprise zone and consequent loss of the dune ecosystem. It was the subject of legal attempts to remove it from listing under the Endangered Species Act so that development could proceed. Kingsley (2002) noted the continuing difficulty of protecting the fly in the circumstances of (1) little remaining habitat; (2) very high value of the land on which it occurs; (3) extremely limited funding; (4) very incomplete knowledge of the fly's biology; and (5) lack of public support for protecting 'a fly'.

Perhaps because many lycaenid butterflies have extremely small distribution ranges and some survive only on small sites that are defined or presumed remnant habitats, a number of them have become important conservation ambassadors both for vulnerable vegetation associations and for specific sites (New 1993), and have helped to promote acknowledgement of insects on wider conservation agendas. They exemplify the 'power' of attractive insects with narrow or remnant distributions, whereas many insects of other orders with similar distributional characteristics but far less 'charisma' have received much less attention – often because they have lacked specific advocacy. Three lycaenids from different parts of the world are noted here. The El Segundo blue (*Euphilotes bernardino allyni*) became notorious because of its presence on the property of Los Angeles International Airport. The South African Brenton blue (*Orachrysops niobe*) and the Australian Eltham copper (*Paralucia pyrodiscus lucida*) were both discovered on small remnant habitat patches scheduled for imminent housing development – the common names reflect the local concerns, 'Brenton' for Brenton-on-Sea, near Knysna on the southern Cape coast, and 'Eltham' being the suburb of outer Melbourne where the subspecies was found, having been believed widely to have been rendered extinct in the region. Such epithets can become a source of considerable local pride and important in promoting local support and awareness. The history of these three cases is summarised by Mattoni (1993), Steencamp and Stein (1999) and Braby *et al.* (1999), respectively.

For the El Segundo blue, the importance of 'listing' was demonstrable. It was remarked by Mattoni (1993) as 'Listing of the El Segundo Blue under the federal Endangered Species Act provides one of the greatest success stories of that legislation'. The listing (in 1976) happened just in time to halt a plan to develop the entire 120 ha of dunes on which the butterfly occurred, and led to setting aside 80 ha of the highest quality land as a permanent reserve. Substantial funding was provided by the Airport Commission to commence restoration of the dune ecosystem. However, one other conservation measure proposed, although very well-intentioned, merits comment here: Mattoni noted that one company involved in the proposed development (Chevron) emphasised creating conditions to maximise butterfly survival at the expense of the ecosystem – through creating a near monoculture of the larval food plant as a 'butterfly garden', with enhanced advertising potential for them. Eventually, and to Chevron's credit, this rather extreme step was not permitted and the area was preserved, as above, so allowing the buttefly to persist at relatively normal density and dispersion.

The Brenton blue had not been seen for more than a century when it was rediscovered in 1977 on the southern Cape Coast, and was signalled as 'vulnerable' in the South African Red Data Book (Henning & Henning 1989). By 1993 it seemed clear that the last remaining population, at Brenton-on-Sea, could be eliminated by holiday house development, and this situation led to a major conservation 'battle' that lasted for more than four years. The decision not to develop the site was not taken until 1997, and the site was later declared a butterfly reserve. Negotiations over that period were very complex. As Steencamp and Stein (1999) commented 'The Brenton Blue became a case study because of its relevance to certain interest groups, including the environmental conservation lobby, and its usefulness as a political tool'. A later evaluation had the main objective to 'use the Brenton Blue issue as a case study to test and inform policy on biological diversity and to increase awareness on the issue among decision-makers and role-players in the environmental field'. The project thus grew from one of initial local interest to a nationally known campaign, and the Steencamp and Stein report is a sobering account of how disparate views by strongly expressed interests may both hinder and facilitate conservation. The area involved at Brenton-on-Sea was only about 2–3 ha.

Habitat patches of similarly small extent are also the continuing focus of conservation interest for the Eltham copper near Melbourne, Victoria, initially also for their impending loss from housing development. The butterfly was believed widely to have become extinct in that region, when a thriving colony was discovered on a small site scheduled for immediate subdivision and development. This discovery occurred at the time (1987) that the State Government was formulating its major conservation act (the Flora and Fauna Guarantee Act 1988), and the butterfly became an important symbol to test the Government's willingness to include invertebrates in the ambit of that legislation. The butterfly's conservation need was accepted enthusiastically by the local community, and within the next year (following an interim conservation order to halt development and facilitate further investigation of the butterfly) the site was purchased (with funds from government and raised by public donations) and, together with subsequently discovered similarly small patches in the region, became one of Australia's first dedicated 'butterfly reserves' (Braby *et al.* 1999).

More generally, promotion of such insects as 'flagship species' is an important component of advocacy, and one purpose of many recovery plans is to display the intricacies of their conservation need to the wider community. Many of the species discussed here have both benefited from such wider interests and had important wider educational roles – not least in signalling the values of insects, and the problems they face. Thus the *Colophon* stag beetles of South Africa (p. 135) and some birdwing butterflies have helped to raise awareness of the extent and dangers of illegal trade in insects; New Zealand weta (p. 63) have demonstrated powerfully the problems to insects caused by introduced vertebrate predators; and the hornet robberfly (*Asilus crabroniformis*) in Britain has been used to draw attention to the insect communities associated with dung, as a habitat overlooked by many people (Holloway *et al.* 2003a). In a related context, visitors to the Addo Elephant Park (South Africa) are greeted at the park entrance with a sign 'Dung beetles have right-of-way', repeated at intervals throughout the park road system. Many visitors are likely never to have considered the increased vulnerability of the beetles resulting from construction of surfaced roads, which have led to increased elephant traffic and defecation on these open areas. Crushing beetles by driving over dung is a recognised threatening process to the Addo flightless dung beetle *Circellium bacchus*, which is largely endemic to the area.

Publicity for any insect conservation campaign can have novelty value to non–scientists, and such simple messages conveyed by relevant signage can evoke local interests, and perhaps lead to volunteer participation (p. 226) – with the caveat that in some cases drawing attention to an insect species or its site may occasionally not be considered desirable because of endangerment (for example, by drawing attention to the presence of commercially desirable species for which collecting could indeed be a threat) or political sensitivity. Likewise, signage in National Parks or other reserves may convey important conservation messages.

Tropical insects

The examples discussed above are all from the temperate regions of the world. Within the tropics, capability to focus on any insect species to the extent equivalent to the three lycaenid butterflies noted above is usually absent, and the few species for which major conservation efforts have been made have sometimes demanded ingenious and wide-ranging approaches to attract support. Queen Alexandra's birdwing (QAB) was noted earlier (p. 68), and has become a major flagship and 'umbrella species' for wider conservation of primary forests in the Oro Province of Papua New Guinea. QAB is endemic to the province. It is depicted on the provincial flag, and is severely threatened by logging of the primary forests it frequents, and by expansion of the oil-palm industry. It is also one of very few insects to have been a specific focus of a foreign aid programme, with its conservation supported in an international agreement involving Australian aid. The plan, for which a strong biological foundation had been laid by Orsak (1992) and Parsons (1992), sought to incorporate human welfare issues as a central support with the then existing butterfly conservation measures based on attempts at captive breeding/ranching *O. alexandrae* and related species to satisfy collector demand through organised marketing of high quality specimens through a government agency, the Insect Farming and Trading Agency (see Parsons 1992, 1998). Such marketing would necessitate changes to the current CITES listing of the butterfly on Appendix 1. Key elements of the international plan were to provide alternative livelihoods for people who would otherwise benefit from forest losses, and to emphasise the values of sustaining the butterfly and its habitat. The project necessitated provision of these tangible wider benefits. Some of these benefits (such as social developments involving increased nutritional awareness, women's health issues, establishment of kitchen gardens, school science classes, and others) were components of conserving the butterfly, in addition to increasing the security of the forest habitat. The project is currently under the management of a local non-government organisation (Conservation Melanesia) who, following the expiration of the international aid programme, continued to foster 'pride of ownership' of QAB among the local communities.

One of very few tropical butterflies for which detailed biological information has been gathered as a basis for instituting practical conservation is the endemic Mexican swallowtail *Baronia brevicornis*. This species, the only extant member of the subfamily Baroniinae and perhaps the most primitive papilionid in existence, is found in deciduous forests in southern Mexico, where it has a very localised distribution within *Acacia* woodlots (Leon-Cortes *et al.* 2004). Traditional land management practices are an important aspect of maintaining the successional *Acacia* woodlots, and Leon-Cortes *et al.* noted that many other species may also depend for these activities for their survival. The main conservation recommendation was to establish connected networks of *Acacia* patches to reduce isolation of butterfly populations and facilitate successful dispersal. The fact that traditional human interactions are involved in sustaining and increasing habitat may indeed be a positive factor because such activities may be able to be focused within the landscape for the butterfly's benefit.

Baronia was listed as 'Rare' by Collins and Morris (1985). However, QAB is one of four swallowtail species listed by them as 'Endangered'. The Jamaican endemic *Papilio homerus* is another, also listed on CITES Appendix 1 ('Prohibited in international trade'). This spectacular butterfly, the largest New World swallowtail, has been a conservation focus for many years. Its basic biology is well understood (Emmel & Garraway 1990), and studies have continued since that time, largely to fulfil the following objectives for a conservation biology programme suggested by Emmel and Garraway to secure its wellbeing, following substantial habitat destruction for coffee plantations and farmland.

1. Intensive study of population ecology and environmental factors characteristic of habitat areas where it occurs.
2. Precise determination of the critical factors of the life cycle which might contribute to increased survival.
3. Negotiations with the Jamaican government to preserve the montane forest in and around the eastern habitats, and suspend cutting or conifer planting.
4. Investigation of a captive breeding programme.

5. Undertaking an international campaign to raise funds for habitat preservation, including the last significant virgin montane forest in Jamaica.

P. homerus was formerly widespread, but has now been reduced to two small populations. An 'eastern population' occurs at the junction of the Blue Mountains and the John Crow Mountains, and a small 'western population' is found in the Cockpit Country. Recent study of the latter suggests that it may now comprise only around 50 adults (Lehnert 2008), and Lehnert emphasised the potential value of a captive breeding programme for *P. homerus*. This could possibly incorporate interbreeding between individuals from the two populations in an effort to counter any current genetic deterioration within each small population.

Captive breeding is the major conservation thrust for the giant Frégate island tenebrionid beetle (*Polposipus herculaeanus*), an arboreal flightless endemic insect on that small island in the Seychelles. It is listed formally as 'Critically Endangered', and is threatened by predation from introduced brown rats. An *ex situ* population was established at the Zoological Society of London in 1996, and has led to establishment of other captive populations of this long-lived species (Pearce-Kelly *et al.* 2007). Eradication of rats from Frégate has now been accomplished, and future options include re-introduction of the beetle to other Seychelles and nearby islands, in addition to Frégate. As with *Papilio homerus* on Jamaica, *Polposipus* has become a frequent symbol in tourist literature and advertising. Similar publicity for selected other tropical insects may contribute to greater awareness of their parlous state.

The above are simply examples of insects whose conservation has indeed evoked public interest and support in some way. Many others could be cited, particularly from Europe or North America, but the factors involved include local pride in 'ownership', appreciation of biological intricacy, and demonstration of the wider place of the insect in the natural community. Each of these aspects merits fostering. Educational exercises (p. 227) toward this end are useful components of any insect conservation plan.

Summary

1. The generally poor public perceptions of insects are an important counter to conservation interest and can be reduced, at least in part, by (a) their novelty value in conservation and (b) their promotion as

local emblems or 'flagship species': symbols through which to garner support for conservation and as vehicles for wider education on values of insects.

2. Butterflies, in particular (together with dragonflies and some beetles), commonly have a more favourable public image. Many threatened species of these have aroused strong interest and support from local community groups and for some species this interest has been pivotal in successful conservation.

3. Publicity for insects, either locally or by national scientific organisations, continues to be an important facet of practical conservation, particularly in some temperate regions of the world. Species focusing to this refined level is generally not possible in the very diverse tropical regions, but a few tropical butterflies and beetles have indeed been elevated in this way, to become important ambassadors for insect conservation. Exhibits of living insects, sometimes associated directly with conservation breeding programmes, can have substantial educational value.

10 · *Insect management plans for the future*

Introduction: the audience and purpose for insect management plans

Many of the important issues in insect species management have been mentioned earlier, and these all appear in various guises in many species-focused plans. A satisfactory species management plan, for an insect or any other taxon, must contain several essential elements but must also be intelligible as a basis for action and, thus, be realistic and practical in scope. This necessitates clearly stated purposes and definition of the audience for that plan. Burbidge (1996) summarised such requirement as follows. For 'purpose' plans should:

1. Enable conservation work related to a species to be based on and guided by accurate information, as well as focused objectives and actions, and detailed forward estimates of cost.
2. Maximise the probability of recovery and minimise the probability of errors, including errors of omission, that might lead to the species becoming extinct.
3. Allow the public to know what is being done to save the species.

The desired audience should include:

1. Those who have a legal responsibility for nature conservation.
2. Those who will be funding and implementing the plan.
3. Those who want to know what is being done.

As Burbidge (1996) noted, it is likely that many of the audience will not have detailed scientific knowledge of the species involved. And, commonly for insects, potential managers may often lack experience and entomological background. It is thus important to couch all parts of the management plans in simple, clear language – if necessary with accompanying explanation of any obscure technical terms. Recovery plans are also prescriptive once adopted formally, and so require commitment of adequate resources for practical implementation. In short, readers must be able to understand both the content of the plan and the logic of the proposals being made, as well as how those proposals are to be implemented. Many plans are reviewed in draft form through the commissioning agency, and some agencies may have a policy of circulating them widely for public comment. If no such review process exists, it is necessary to arrange it, with a specific request to assess transparency of the proposals as an important aspect of feedback.

Although flexibility is clearly desirable, the constitution of a management or recovery team needs careful consideration, but must incorporate the variety of expertise and advice needed in any particular case. Australia has clear guidelines for membership of a recovery team, which will normally be constituted at an early planning stage and be involved in and responsible for designing the recovery plan. The chair is usually a representative of the lead agency who will implement the plan (Male 1996), and other members may be drawn from people undertaking research or management actions, specialist biologists, a federal Endangered Species Unit representative, funding agency representatives as appropriate, captive breeding institute representative if appropriate, business or local government representatives as appropriate (for example if forestry or mining activities are affected, or municipal developments involved, or for local managerial involvement), and community representatives. This composition is not in any way pre-emptive or prescriptive but is noted here simply to indicate the benefits of the widest possible involvement and expertise at this stage, so that the plan may be operable rather than compromised from the start by unreality, or by serious omissions. In principle, all interested parties should be involved at this early stage, not least for the goodwill this will foster, but the size of the team should not become overly large so as to be unwieldy. The main practical need is to foster effective and informed collaboration between researchers, managers, bureaucrats and the wider community, because all are likely to become involved in a conservation management programme, but may have rather different priorities, background knowledge and points-of-view (Field *et al.* 2007). Because of the predominant need for considerations of plant ecology in many insect conservation exercises, botanical and horticultural expertise can be a core requirement in management planning. Likewise, (1) statistical advice may be needed as a core element in both planning and subsequent analysis of monitoring and other data, not least to assure the quality of any sampling or monitoring programme proposed; and

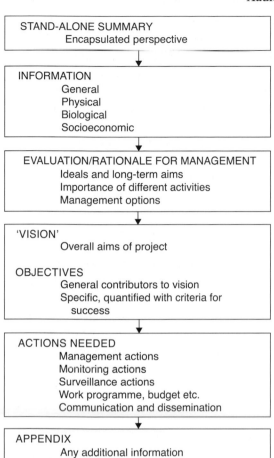

STAND-ALONE SUMMARY
Encapsulated perspective

INFORMATION
General
Physical
Biological
Socioeconomic

EVALUATION/RATIONALE FOR MANAGEMENT
Ideals and long-term aims
Importance of different activities
Management options

'VISION'
Overall aims of project

OBJECTIVES
General contributors to vision
Specific, quantified with criteria for
success

ACTIONS NEEDED
Management actions
Monitoring actions
Surveillance actions
Work programme, budget etc.
Communication and dissemination

APPENDIX
Any additional information

Fig. 10.1. A generalised management scheme as a basis for a species conservation plan (after Ausden 2007).

(2) genetic expertise may be needed if a complex *ex situ* programme is contemplated.

The outline scope or content of any particular recovery plan will be dictated in part by any governing legislation. One possible scheme for planning content is indicated in Fig. 10.1. Many variations on this are possible. However, a number of 'working guidelines' may be suggested (Burbidge 1996), so as to incorporate a balance between reality (based on knowledge and logistic capability) and idealistic SMART aims. It is important not to consider interim plans, or those based on inadequate biological knowledge, as definitive, as has sometimes occurred without provision for effective review. Points to consider for inclusion vary, but could include careful thought on the amount of basic background information to provide. For the major audiences (above), Burbidge suggested that this information should be 'brief and to the point'. Plans with comprehensive background are valuable reference documents, but much of the detail may not be strictly relevant to the desired outcomes, and its inclusion may provide unnecessary complications for readers. In such cases, supporting information may be placed better as auxiliary documentation, which can be easily available if needed during the recovery process, or as appendices to the core plan. Objectives, and all actions to achieve these, should be clear, measurable, and generally SMART. They should be costed as accurately as possible, while recognising that changes will inevitably occur. Prescribed recovery actions are central to any plan, and the programme will succeed or fail largely on the clarity and feasibility of these. Clear statements on actions should be accompanied by designation of duties to a particular agency or other person for implementation, with sources of funding identified. A relevant caveat here is that funding requirements should not be overstated beyond projections for inflation. Inclusion of funding requests for 'tangential' aspects rather than the core needs can deprive other species of those funds, and it is important not to inflate the 'research component' to include matters not strictly related to undertaking management or recovery. Such 'wish lists', however tempting and easily justifiable as 'needed strategic knowledge', are not part of a formal management plan.

The above prescription does not mean that preparation for a management plan should ignore *any* available information. Clearly, comprehensive gathering and review of *all* available data on the species, and on any specific sites targeted for management must be undertaken. However, it is often not necessary to report all fine detail in the final plan, as long as users are made aware of its availability if needed, and have confidence in the scientific robustness and common sense of the plan. From the management viewpoint, much basic 'research' on a threatened species is not necessarily a focused contribution to conservation, other than by adding to the corpus of basic biology that might eventually facilitate greater understanding. Part of the value of a good recovery plan is to reveal the specific research questions through which practical conservation management may be addressed and enhanced.

Wherever possible, plans should promote and facilitate public/ volunteer/community participation. Linked with this, funding from the public purse, with due acknowledgments to sponsors and community groups (as appropriate), is a vital component of many management actions, for which success depends largely on continuing goodwill. Many of the plans noted earlier could not have proceeded without such help and support as a major factor in management, over a wide variety of activities. Nevertheless, effective central coordination is vital. As Nally (2003) put it for *Paralucia spinifera* 'The conservation of threatened species through community engagement requires an interdisciplinary approach, allowing groups to incorporate recovery actions into group activities, or to choose their method of contribution'. Undertaking a long-term community-based recovery programme for an insect is sometimes feasible, but continued support demands a high level of awareness, tact and effective liaison through practical involvement, rather than simply directives from 'remote' scientists or policy makers. Various means exist to help assure continued interest and to help 'recruit' further people. Many insect conservation projects issue occasional newsletters to community groups, and these may include matters such as protocols for food plant propagation and planting, notices of meetings or talks of relevance, profiles of key people in the group, notes on key advances or the insects involved, and so on. Community group meetings, sometimes on field sites as adjuncts to working bees, or with visiting speakers, can also be popular and effective. And family social 'thank-you' events (involving young people and their older relatives) such as barbecues or picnics demonstrate appreciation of volunteer efforts effectively.

The spirit of the last paragraph encapsulates the more formal needs for promoting community interest in species management reviewed by Williams (1996); although her experiences were drawn from birds, the principles extend easily to insect conservation. Williams suggested that government agencies (extended here to other 'controllers', including recovery teams) should provide community endeavours with 'honesty, support, expertise and a sensitivity to the community's concerns for conservation', through the factors listed in Table 10.1. Conversely, some factors can alienate community interest rapidly, and failure to acknowledge these adequately can be highly detrimental. Nally (2003) noted that failure of the community to identify with a site, or to recognise direct benefit of any conservation action to the community (such as by lack of opportunity to express personal viewpoints, imposition of unrealistic or inadequately explained or understood viewpoints, and request for work beyond the community group's capacity: Williams 1996), led to lack of volunteer interest.

Table 10.1 *Points to help foster community interest and involvement in species conservation programmes*

1. Have a focus for conservation interest with which the community or community group identifies personally (effective focus)
2. Encourage community involvement from the earliest developmental stages of a conservation or species recovery initiative (sense of ownership)
3. Develop programmes that are beneficial to the community as well as to conservation (what does the community 'gain' from the exercise and effort?)
4. Listen to the community's concerns (constructively incorporate them into the conservation goals)
5. Gain the community's trust (personal interactions and considerations important)
6. Provide the community with the appropriate information at the appropriate level and at the appropriate time (regular review and feedback; effective communication)

Source: after Williams (1996).

At an early stage, it is worth reflecting on the five major 'tasks' needed to solve any problem through a strategy such as a recovery plan, and which can help initial orientation and focus. It is given that we have an insect species selected for conservation management. Clark (1996) listed these initial considerations as follows.

1. Clarification of goals. This demands defining the preferred outcomes in terms of the 'states' (such as population size, number of populations) to be realised, and defining the problems of achieving these, together with the principles and practices of solving those problems.
2. Trend descriptions. Defining the extent to which past and recent events have approximated the desired outcomes, and evaluating extent and significance of discrepancies.
3. Analysis of conditions. Defining what factors have affected (or conditioned) or caused the direction and magnitude of those trends.
4. Projection of developments. Assessing the probability of attaining the goals if current policies are continued.
5. Invention, evolution and selection of alternatives. Defining which alternatives (research, practical management, innovation) will help to realise the goals.

These may be tabulated (following Clark 1996 and Lasswell 1971) in the form of a scenario for the species, in which the specific points that arise can be addressed, as in Table 10.2.

Table 10.2 *A problem-oriented exercise on the threatened species recovery process*

The questions are posed, but the answers are for the practitioners (recovery teams) to appraise for the particular taxon/taxa involved.

Problem-solving task	Application
1. Goals. What are the preferred outcomes?	Species sustainability, recovery, enhancement? (etc.)
What are the problems[a] with respect to goals?	Problems: principles and practice; sufficiently clear; consensual?
2. Alternatives. What alternatives are available to solve the problems?	Science-based, practice-based, innovation-based?
3. Evaluation of alternatives. Would each contribute to solving the problem? Trends. Did it work when tried in the past, or on other species? Conditions. Why, and under what conditions, did it work or not work? Projections. Will it work satisfactorily in this case?	Which, how?
4. Report procedure to refine and update goals, alternatives and evaluation.	Adaptive management

[a] Problems are discrepancies between goals and actual or anticipated states.
Source: after Clark (1996).

The above, or a similar set of pointers, are useful guides for recovery plans, when the data from monitoring can be used to indicate trends toward defined goals, any unexpected or novel outcomes, and the need for change or additional measures. In some cases, only selected steps from a portfolio of management actions may be undertaken initially, with others delayed deliberately as less urgent, regarded as optional or alternative depending on outcomes of other actions, or awaiting the results of targeted research activities.

Constructing an insect recovery plan

With the above considerations in mind, together with the proforma schemes noted earlier, we can consider the optimal procedure and components for an insect recovery plan more fully, to reflect both the general needs and the more specific considerations involved. One possible sequence is as follows.

1. Select the taxon for treatment, on the basis of need (current or pro-jected) for conservation based on threat and demonstrated declines or losses leading to consensus over priority. It may be necessary to ensure (1) that the taxon is a valid taxonomic entity, borne out by specialist taxonomic advice, or otherwise a significant evolutionary entity, and (2) that it is acknowledged formally as worthy by some designation on a legal schedule, as a condition of government or other agency support. Selection may also flow from acceptance of a formal nomination for 'listing', in which major perceived conserva-tion status and needs are included. The taxon must both be worthy of conservation focus, and be seen to be worthy.

2. Appoint or gather an interim recovery team, with representatives from all interested parties and adequate scientific expertise, to be involved in assessing conservation needs and drafting the recovery plan.

3. From this membership, accumulate and review all ecological infor-mation (including historical information) on (1) the species/other taxon/other entity targeted, (2) the rationale for its conservation (including information on all current and anticipated threats), and (3) the environment (current and future) and security of sites on which conservation is projected. The last of these should not be neglected, and may involve title searches and effective liaison with public bodies and local government. Information should be gathered from all parts of the taxon's range, not just the local environment of concern, and particular attention given to any other conservation concerns or plans that exist. In some instances, coordination with planners for the same species in other parts of its 'political range' may be needed. Cross-membership of recovery teams can then be valuable.

4. From that information, with additional background from related species or other taxa on the same sites, assess and define the major components of conservation need, and draft objectives for incorpo-ration in the plan. If necessary, seek peer review of the worth and feasibility of those objectives, together with seeking consensus within the recovery team and wider community. At this stage, or before, valuable additional perspective may be gained from discussions with people who have raised concerns about the insect (for example, as the nominators for listing) if they are not involved already.

5. Decide whether the information available from steps 3 and 4 is adequate to proceed.

(a) If so, clarify objectives, and nominate key actions needed in SMART terms (go to 7).

(b) If not, define the key areas for which more information is needed, and explore how that information may be obtained by targeted research, within a given time frame (go to 6).

(c) In conjunction with 'b', determine the key urgent management steps needed to 'preserve the system' during this research interval, and to stop any ongoing losses or declines (go to 6).

(d) Initiate education and publicity as deemed wise to broaden support for the conservation programme.

(e) Possibly augment the interim recovery team to a more definitive constitution from previously unknown 'players' detected during steps 3–5.

6. Once outcomes are clear, review and revert to 5a; if necessary, 're-loop' though 5b and 5c.

7. Formulate all actions in SMART terms, with indicative funding and designation of lead agency or other responsible personnel to undertake and supervise each action proposed.

8. Integrate all objectives and actions into a cohesive flow scheme, with 'step-down' actions specified and allocated as above, and refine costings.

9. Designate priorities amongst these objectives and actions, and determine if any can be pursued in concert for greater efficiency and economy. At this stage, list and consider alternative management steps/needs that might come into play at some later stage: as far as possible ensure that important options have not been overlooked.

10. Define milestones and review dates/stages. Define review process(es), and consider alternative management strategies well in advance of review. Ensure provision for central deposition and storage of all archival and newly acquired information relating to the plan.

This exercise, or some equivalent disciplined protocol, is in many ways the easier phase of a recovery plan in practical conservation. Bringing the plan to fruition is usually more difficult because, despite the most thorough planning, 'things' rarely go as planned! Funding and other logistic support may not be forthcoming as anticipated or needed, site security proves less than anticipated, unusual weather hampers access or sampling, insects die in captivity or do not breed, stochastic events intervene, and so on. It follows that the 'alternatives' suggested in the above scheme can become very important, and that some form of time flexibility and

financial buffer should be included wherever possible, notwithstanding the caveat noted earlier. Such uncertainties emphasise further the
importance of reviewing the plan, perhaps first after a relatively short
period of operation such as one field season, with the aims of detecting any gaps in coverage, inadequacies of current management due to
unexpected events, and logistical difficulties revealed at this stage. Later
review is likely to include also evaluating reasons for interim success
or failure. Indeed, recovery criteria may need first to be defined as
'interim', because further research is needed in order to make them fully
measurable. Thus, as for Hungerford's crawling water beetle (*Brychius
hungerfordi*) in the United States, an interim suite of tasks to make criteria
measurable was needed as a prelude to defining fully usable advice (Tansy
2006).

Major components such as the above do appear in many insect recovery plans, of course, but they are often not explicit, and commonly lack
detail. The ecological background available on the species largely determines the balance between research and management; the sites/habitats
involved guide the need for formal measures for site protection and for
additional agreements with landowners or managers; the extent and form
of external threat(s) may guide ways for effective advocacy and education (as well as for assessing the role of the insect in wider conservation
needs and promoting collaborative programmes); and the need for formal
recognition (listing) of the species will reflect the local political environment in relation to funding opportunities. One recovery consideration
must inevitably be to ensure that specific habitat or resource needs are
adequately incorporated in wider conservation agendas for the same sites,
if those sites are the focus also of management for other taxa. It is also
useful practice to allocate a level of 'priority' and 'feasibility' to every
action proposed, as valuable guides to reality and honing actions to being
SMART. Priority can be adjudged on a simple scale – such as 'very
high', 'high', 'moderate' and 'low', or an equivalent number sequence,
but 'feasibility' may command deeper thought. Lundie-Jenkins and Payne
(2000) used a percentage scale in their plan for *Hypochrysops piceatus* in
Queensland, and a general guide could be to ensure that all actions are at
least 80% feasible, and preferably 100% so. If not, either (1) the means to
render them more feasible should be specified and assessed, or (2) they
should be discarded and replaced by others. Occasionally, obstacles to
high feasibility are 'political' rather than scientific and in such cases lower
levels of feasibility may have to be accepted temporarily, while the first
of the above alternatives proceeds.

As one common example of low feasibility, development of population viability models to investigate population size and structure is recommended in a number of recovery plans (such as that for Hine's emerald dragonfly in North America: USFWS 2005). This may be a basis for reviewing recovery criteria or to determine whether a naturally changing distribution provides measures equivalent to recovery criteria. Such information can indeed be of very considerable value but, for most species, is a luxury that may eventuate in the future but cannot be a prerequisite for current conservation planning. Nevertheless, monitoring data can be accumulated with the aim of providing the foundation for this in the future. Detailed population data, as noted earlier, are one of the almost universal lacunae in insect conservation assessment. Any monitoring of recovery ventures and allied management may contribute constructively to redressing this lack, and this purpose merits consideration in designing the requisite proformas. Allied with estimation of population sizes, debate will assuredly continue over numerical thresholds (as in the IUCN categories of threat, p. 7) but, as Clarke and Spier (2003) noted, the most serious problem is interpreting numerical changes caused by key threats against a background of normal large fluctuations from generation to generation or year to year. This dictates that long series of observations may be needed to detect any trend due to a threat or, conversely, to reflect success of management. Many related concerns persist (Hutchings & Ponder 1999).

Budget planning can be helped considerably by constructing a flow chart of actions over the duration of the plan or, at least, for the initial period of its implementation, and which also reveals how the priority individual steps can be orchestrated and integrated. Insect species recovery must be viewed as a long-term process. In Australia, the community-headed recovery programme for *Ornithoptera richmondia* (p. 122) has been in train for more than 20 years, with some notable successes. In addition to the wide community interests noted earlier, it was selected by Environment Australia for showcasing to overseas media during the 2000 Sydney Olympic Games, for example.

Alternatives to a formal management plan?

For species that lack a formal management plan, it is often necessary to introduce a suite of conservation objectives on a less formal basis, simply to ensure that conservation can be promoted and proceed in the best possible interim manner while such a plan is being devised.

Table 10.3 *Research and management for conservation of the Karner blue butterfly in New York State: the spectrum of activities flowing from 'operational goals'*

Extensive and intensive surveys for Karner blue butterflies and lupins.
Estimation of population size at selected sites (mark-release-recapture).
Annual monitoring of populations and habitats, and determination of population trends through index (walk through) counts.
Creation and expansion of suitable habitat, specifically lupin, through studies of lupin germination and vegetation control techniques.
Development and implementation of specific management plans for selected sites.
Establishment of cooperative agreements with key landowners for select Karner blue butterfly sites.
Reviews of project proposals impinging on Karner blue butterfly or lupin habitats, and recommendation of mitigation actions.
Preparation and dissemination of information and educational materials pertaining to Karner blue butterfly.
Direct notification of town governments of locations of Karner blue butterflies, and requests for them to cooperate in site protection and reviews.

Source: Sommers and Nye (1994).

Conservation might proceed according to a set of 'operating goals' that can each be backed by actions. This has been done for the Karner blue butterfly in New York (Sommers & Nye 1994), where the goals (aimed at survival of the butterfly) are:

1. No loss or decline of occupied Karner blue habitat.
2. Stabilisation of existing occupied habitats and populations.
3. Augmentation (expansion) and intensive management of the three or four most significant population sites.
4. Re-creation of lupin corridors and dispersal pathways adjacent to and between occupied sites.
5. Translocation of butterflies into occupied areas.
6. Full establishment of at least three metapopulations within New York State.

The butterfly has been designated officially as 'Endangered' in New York State since 1977, so that the above objectives could incorporate considerable biological understanding. That might not always be possible. Research and management actions for the Karner blue flowing from the above list and based on the demands of the State's endangered species programme are noted in Table 10.3.

Consultation and acceptance

Each of the steps above may involve some level of innovation or unexpected occurrences. Thus, many insects have been catapulted to conservation significance as a consequence of planning decisions to develop the site(s) on which they exist, and plans for a supermarket, freeway, housing estate, exotic tree plantation, golf course (and so on) may be well advanced, and even contracts let to developers, before the potential crisis is revealed. In some cases, little can be done to forestall these but various avenues may be open in others, together with local 'weight' from species listing. There are notable cases of major developments being halted or diverted to accommodate sites for threatened insects. Several were noted earlier, but a major planning need is to try to anticipate such events if sufficient advance warning is given and, perhaps, to incorporate legal expertise on the recovery team if any such need is present. Likewise a representative of any sympathetic developer may be helpful for mutual benefit and some compromise in the design of any focused or wider environmental impact statement required for such sites, or some form of 'offset' through which an alternative site might become available for the insect. Not in any way restricted to insects, options such as conservation easements or agreements may be available to help assure site security for the longer term. Many forms of 'agri-environment schemes' exist in various places, and can often be a valuable adjunct to insect species management. Wherever the focal insect occurs within or near agricultural landscapes, opportunities related to any such schemes operating in the region should be explored in relation to increasing site security and the range of management options. More broadly, much insect conservation takes place on private lands, and several guidelines are needed to ensure that this can occur successfully. Each case will differ, of course, but a minimum suggestion for fostering goodwill could comprise (1) contacting private landowners individually to explain the need and ascertain their feelings; (2) providing management guidelines to private landowners and, where possible, offer practical or financial support or compensation for those activities; and (3) promote landowner involvement in site restoration and protection, perhaps as a member of the management group. Possible incentives for private landowners vary widely in different places, but may include some forms of 'biodiversity credit' or other subsidy for conservation activity. The recovery team may need to explore all of the possibilities formally within their area of interest.

An allied need is to consult effectively with traditional landowners, where these occur, and to ensure their representation in planning and practical management. Thus, New Zealand recognises formally the need for consultation with Maori, with the process recognising that researchers (specifically in biological control in the context cited) may be required to step outside their usual perspective to take into account the spiritual values, perspectives and opinions of a culture vastly different from the familiar 'Western scientific culture'. In such cases 'Consulting involves the statement of a proposal not yet finally decided upon, listening to what others have to say, considering their responses, and then deciding what will be done' (McGechan in BIREA 2007), and further guidelines exist on what constitutes 'effective consultation'.

Liaison with formal community groups, such as local environment, conservation or natural history societies may lead to offers of support and help, as well as aiding advocacy and goodwill. Support may be forthcoming for specific aspects of management, including volunteer labour for site restoration, planting exercises, participation in surveys and other components. 'Working bees' and the like are valuable exercises to pursue, and in principle the list of activities that could be involved is long. It is useful to prepare a list of possibilities well in advance – should any unexpected offer of help then arise, perhaps as a result of publicity or signage, a rapid and positive response from the recovery team can pay dividends. A specified recovery team member might take on the role of 'community liaison officer', as an initial contact point for interest. Very few insect management proformas spell out such possible community contributions in advance. However, for the Mardon skipper butterfly (*Polites mardoni*), a threatened grassland species in the northwestern United States, BCI (2006) listed a number of educational activities in which local school and community groups might participate. In conjunction with this, they noted the considerable values of local examples of threatened species for curricular studies that could be fostered by site visits and meetings with the scientists involved. BCI's list included:

1. Studying the skipper (or other insect) life stages.
2. Researching the habitat needs of the skipper.
3. Corresponding or meeting with the biologists managing current skipper sites.
4. Visiting sites during the adult flight season, to see the skipper.
5. Visiting captive breeding programmes.

6. Assisting with on-site management.
7. Propagating and growing host plants for planting at butterfly sites or for use in captive breeding programmes.
8. Writing letters to decision makers to ensure that the skipper receives adequate resources and protection.

Many of these points recur in practice in insect recovery programmes, and may enter educational curricula in various ways. Cherry Lake, near Melbourne (Australia) is a stronghold for the Altona skipper butterfly (p. 227) and is the focus of a broad educational programme on the values, environment and conservation of wetlands (DNRE 2001). Sequences of classroom lessons are complemented by structured exercises based on class visits to the site. Although not specifically directed at *Hesperilla flavescens*, exercises such as this create lasting awareness in young people of problems in their local environment and, in this instance, may help to guide some to a keener interest in insects, environments and conservation. Schools' participation throughout the species' range has long been a keystone of the Richmond birdwing butterfly (p. 61) conservation plan (Scott 2002). That exercise started on a small scale but grew rapidly following its launch in 1993, with more than 130 schools becoming involved within the next year. Targeted educational activities may prove valuable adjuncts in many such programmes. Some North American recovery plans use the term 'outreach program', with the intent of keeping local communities informed of the insect's status. The Richmond birdwing programme includes suggestion for development of interactive school packages for younger children (Siepen 2007). As examples of the wide-ranging involvement that may be developed within primary school curricula, Siepen listed the following activities that may be involved.

1. Watering and caring for vine seedlings in the shade house, and learning about the basic care requirements to cultivate plants before they are planted out (for Science)
2. Measurements of growth rates, water use, etc. (Mathematics)
3. Potting up soil mixes for future plantings and using different fertilisers (Science)
4. Learning about the various life stages of the butterfly and its relationships with the host vine (Reading, English)
5. Visiting sites where vines grow naturally or have been planted (Social Sciences, Science)
6. Planting vines in suitable sites in the school grounds (Science)

7. Paintings and/or illustrations of the plant and the butterfly (Arts)
8. Reading stories about similar plant/animal relationships (English)
9. Life cycle depictions (Arts)
10. Learning where the vine/butterfly used to live and why they have declined (Social Sciences, Science).

The emphasis of any such programme is to be 'hands-on' and interactive. Similar approaches may be possible for many insects of conservation concern. However, activities that may have adverse effects, such as causing damage or interference on small sites, must be avoided, so activities as diverse as those suggested above may not always be feasible.

The need for 'adaptive management' has been stressed repeatedly, to ensure that a plan does not fossilise or lose touch with the real world. The basis for this largely comprises effective monitoring and review, so that the parameters for this need to be defined at an early stage. A review may be 'external' in calling for information from the wider community (in which case appropriate advertisement and a lead-in period are needed), or be based mainly on appraisal by the recovery team and, perhaps, additional invitees. In the former case, some authorities request documentation to confirm new information. Review may simply call for information and advice, or may be couched in responses to a particular carefully formulated set of questions addressing outcomes of the objectives and actions projected for progress by the review date. Wider parameters may be relevant, based on new knowledge and how this may change perception of an insect's status and needs. Under the United States Endangered Species Act, for example, a five year review of a listed species is 'assessment of the best scientific and commercial data available at the time of review', and this is listed under five headings:

1. Species biology, including but not limited to, population trends, distribution, abundance, demographics and genetics;
2. Habitat conditions, including but not limited to amount, distribution and suitability;
3. Conservation measures that have been implemented to benefit the species;
4. Threat status and trends (against stated criteria of habitat destruction or change; overexploitation; disease or predation; inadequacy of existing regulatory measures; other natural or manmade factors affecting its continued existence);

5. Other new information, data or corrections, including but not limited to taxonomic or nomenclatural changes, identification of erroneous information in the earlier listing, and improved analytical methods.

Collectively, a range of directions such as the above encompasses much of the need for updating a plan by evaluating the status of a species and its environment based on the widest appraisal available. For evaluating the directions of a management plan, the third of the headings may be a central consideration because, in a very high proportion of cases involving insects, continuing conservation measures will indeed be needed, and several of the other categories above are the major monitoring indices against which success (or otherwise) of this may be appraised.

After recovery?

The objectives of many insect recovery plans include a statement of intent to increase the security of the species so that it can be removed from a protected species list as no longer in need of conservation management. An initial appraisal may be needed at the planning stage of management to assess the ease and procedures by which this may be accomplished, and the criteria defined to mark when this might occur. For some legislations, the act of 'de-listing' is difficult, with authorities tending to regard listing as a permanent state. In other contexts, de-listing is the major desired outcome of the initial listing step, and can be undertaken easily once recovery objectives (measured against agreed criteria) have been achieved. More broadly, five reasons for de-listing an insect may occur, and each such removal of a taxon from a priority list may release resources that can then be devoted to other, now more needy, taxa. The two major reasons for de-listing are:

1. When additional survey or research prompted or facilitated by the action of listing reveals the insect to be more secure (more widely distributed, in greater numbers, less threatened) than believed initially, so that it does not currently need conservation management.
2. When recovery actions (undertaken against SMART criteria) have been successful and the insect is considered to be out of danger.

The first of these is perhaps the more common outcome at present but, for most such species, the requisite study would never have been undertaken without the initial listing, and the value of the precautionary principle in initially listing the taxon in cases of uncertainty can be considerable.

The other three contexts are noted here mainly for completeness, but may occasionally become important:

3. When there is no reasonable doubt that the focal insect has become extinct, both in the wild and in captivity.
4. When taxonomic revision reveals that what was believed to be a distinctive taxon is, in fact, not so. Some caution is needed here, because of the importance of 'significant populations' and geographically distinct 'forms' (see the Altona skipper butterfly, p. 38), and some legislations may need modification in formal scope to accommodate such cases more easily.
5. As suggested for British insects by Key *et al.* (2000), but not yet flagged widely, when successive attempts to implement conservation measures all come to nothing, and consequently the species is deemed to be beyond recovery.

The second of these five categories is the most important and informative for future endeavours. Sands and New (2002; see also New & Sands 2004) suggested that such taxa should be recognised as 'rehabilitated species', in which conservation investment had been successful and should not be forgotten. Such species merit monitoring into the future to ensure that their circumstances do not deteriorate to again threaten them. Some level of stated formal value of these species may ensure their future protection by protecting the earlier conservation effort, and also provide long-term monitoring data to add to understanding of those species.

Summary

1. Insect conservation management plans must inform three major categories of people: those with policy or legal responsibility for conservation, those responsible for funding and implementing the plan, and those interested in the exercise and who seek knowledge on 'what is being done'. Plans must be couched in very clear terms, and contain adequate but not overwhelming detail.
2. Plans are prescriptive, and may be legally binding, so that readers must be able to understand the objectives and actions, and how the plan is to be fulfilled in practice.
3. An initial 'recovery team' to oversee design of the plan and its implementation needs should include a wide spectrum of relevant expertise and interest to give a comprehensive and informed perspective, to ensure that important matters have not been overlooked, and (as far

as possible) also to ensure that the plan is couched in SMART terms. Effective communication and education are essential components of many insect conservation plans, and community or hobbyist support and involvement depends on trust and respect.

4. Construction of a good insect species management plan is a complex exercise, and a tentative proforma process based on a sequence of ten steps is outlined and discussed, to illustrate some of the matters that need consideration. Periodic review of the plan is necessary, so that management projected must be flexible and responsive to findings and trends in the future. For many insect plans, a possible formal outcome is declaring 'recovery' of the species, so that it no longer needs to be listed formally as threatened. Such de-listed taxa represent substantial conservation investment, and can be recognised as 'rehabilitated species'.

References

Abbott, I., Burbidge, T. & Wills, A. (2007). *Austromerope poultoni* (Insecta, Mecoptera) in south-west Western Australia: occurrence, modelled geographical distribution, and phenology. *Journal of the Royal Society of Western Australia* **90**, 97–106.

ACT Government (1998). *Golden Sun Moth* (Synemon plana): *an Endangered Species.* Action Plan no. 7. Canberra: Environment ACT.

Ahern, L. D., Tsyrlin, E. & Myers, R. (2003). *Mount Donna Buang Wingless Stonefly* Riekoperla darlingtoni. Action Statement No. 125, Flora and Fauna Guarantee Act 1988. Victoria: Department of Sustainability and Environment.

Akçakaya, H. R. & Ferson, S. (1999). RAMAS® Red list: Threatened species classification under uncertainty. Version 1. Setaulcet, NY: Applied Biomathematics.

Albanese, G., Nelson, M. W., Vickery, P. D. & Sievert, P. R. (2007). Larval feeding behaviour and ant association in frosted elfin, *Callophrys irus* (Lycaenidae). *Journal of the Lepidopterists' Society* **61**, 61–6.

Albanese, G., Vickery, P. D. & Sievert, P. R. (2008). Microhabitat use by larvae and females of a rare barrens butterfly, frosted elfin (*Callophrys irus*). *Journal of Insect Conservation* **12**, 603–15.

Andrew, N. R. & Hughes, L. (2004). Species diversity and structure of phytophagous beetle assemblages along a latitudinal gradient: predicting the potential impacts of climate change. *Ecological Entomology* **29**, 527–42.

Andrew, N. R. & Hughes, L. (2005). Diversity and assemblage structure of phytophagous Hemiptera along a latitudinal gradient: predicting the potential impacts of climate change. *Global Ecology and Biogeography* **14**, 249–62.

Anon. (1996). *Papua New Guinea Conservation Project. Oro Province. Project Implementation Document.* Canberra: Papua New Guinea Department of Environment and Conservation and Australian Agency for International Development.

Anthes, N., Fartmann, T., Hermann, G. & Kaule, G. (2003). Combining larval habitat quality and metapopulation structure – the key for successful management of pre-alpine *Euphydryas aurinia* colonies. *Journal of Insect Conservation* **7**, 175–85.

Anthes, N., Fartmann, T. & Hermann, G. (2008). The Duke of Burgundy butterfly and its dukedom: larval niche variation in *Hamearis lucina* across central Europe. *Journal of Insect Conservation* **12**, 3–14.

Arnold, R. A. (1983a). Conservation and management of the endangered Smith's blue butterfly *Euphilotes enoptes smithi* (Lepidoptera: Lycaenidae). *Journal of Research on the Lepidoptera* **22**, 135–53.

Arnold, R. A. (1983b). Ecological studies of six endangered butterflies (Lepidoptera: Lycaenidae): island biogeography, patch dynamics and the design of habitat preserves. *University of California Publications in Entomology* **99**, 1–161.

Asher, J., Warren, M., Fox, R. *et al.* (2001). *The Millennium Atlas of Butterflies in Britain and Ireland*. Oxford: Oxford University Press.

AusAID (1999). *Papua New Guinea Conservation Project. Oro Province. Project Completion Report*. Canberra: Australian Agency for International Development.

Ausden, M. (2007). *Habitat Management for Conservation. A Handbook of Techniques*. Oxford: Oxford University Press.

Bandai, K. (1996). Preserving woodland for butterflies – conservation of the great purple, *Sasakia charonda*, in Nagasaka-cho, Yamanashi Prefecture. *Decline and Conservation of Butterflies in Japan* **3**, 180–4.

Barnett, L. K. & Warren, M. S. (1995). *Species Action Plan. Silver-spotted Skipper* Hesperia comma. Colchester: Butterfly Conservation.

Barratt, B. I. P. (2007). Conservation status of *Prodontria* (Coleoptera: Scarabaeidae) in New Zealand. *Journal of Insect Conservation* **11**, 19–27.

BC (Butterfly Conservation) (1995). Lepidoptera restoration: Butterfly Conservation's policy, code of practice and guidelines for action. *Butterfly Conservation News* **60**, 20–1.

BCI (Butterfly Conservation Initiative) (2006). Mardon skipper butterfly *Polites mardoni*. www.butterflyrecovery.org/species_profiles/mardon_skipper/

BDS (British Dragonfly Society) (1988). *Pond Construction for Dragonflies*. Purley: British Dragonfly Society.

Beale, J. P. (1998). Comments on the efficacy of Queensland nature conservation legislation in relation to *Acrodipsas illidgei* (Waterhouse and Lyell) (Lepidoptera: Lycaenidae: Theclinae). *Pacific Conservation Biology* **3**, 392–6.

Beaumont, L. J. & Hughes, L. (2002). Potential changes in the distributions of latitudinally restricted Australian butterfly species in response to climate change. *Global Change Biology* **8**, 954–71.

Beaumont, L. J., Hughes, L. & Poulsen, M. (2005). Predicting species distributions: use of climate parameters in BIOCLIM and its impact on prediction of species' current and future distributions. *Ecological Modelling* **186**, 250–69.

Begon, M. (1979). *Investigating Animal Abundance: Capture-Recapture for Biologists*. London: Edward Arnold.

Beynon, T. G. & Daguet, C. (2005). Creation of a large pond for colonization by white-faced darter *Leucorrhinia dubia* dragonflies at Chartley Moss NNR, Staffordshire, England. *Conservation Evidence* **2**, 135–6.

Binzenhöfer, B., Schröder, B., Strauss, B., Biedermann, R. & Settele, J. (2005). Habitat models and habitat connectivity analysis for butterflies and burnet moths – the example of *Zygaena carniolica* and *Coenonympha arcania*. *Biological Conservation* **126**, 245–59.

BIREA (2007). (Biocontrol Information Resource for ERMA NZ Applicants), quotation from Mr Justice McGechan. www.b3nz.org/birea/index.php/page=background_maori_nature, accessed 18 July 2007.

Boersma, P. D., Kareiva, P., Fagan, W. F., Clark, J. A. & Hoekstra, J. M. (2001). How good are endangered species recovery plans? *BioScience* **51**, 643–9.

Braby, M. F. & Dunford, M. (2007). Field observations on the ecology of the golden sun moth, *Synemon plana* Walker (Lepidoptera: Castniidae). *Australian Entomologist* **33**, 103–10.

Braby, M. F., Van Praagh, B. D. & New, T. R. (1999). The dull copper *Paralucia pyrodiscus*. In Kitching, R. L., Scheermeyer, E., Jones, R. E. & Pierce, N. E. (eds) *Biology of Australian Butterflies*. Melbourne: CSIRO Publishing, pp. 247–60.

Brereton, T. M., Warren, M. S., Roy, D. B. & Stewart, K. (2008). The changing status of the Chalkhill Blue butterfly *Polyommatus coridon* in the UK: the impacts of conservation policies and environmental factors. *Journal of Insect Conservation* **12**, 629–38.

Britton, D. R., New, T. R. & Jelinek, A. (1995). Rare Lepidoptera at Mount Piper, Victoria: the role of a threatened butterfly community in advancing understanding of insect conservation. *Journal of the Lepidopterists' Society* **49**, 97–113.

Brower, L. P. (1996). Forest thinning increases monarch butterfly mortality by altering the microclimate of the overwintering sites in Mexico. *Decline and Conservation of Butterflies in Japan* **3**, 33–44.

Burbidge, A. A. (1996). Essentials of a good recovery plan. In Stephens, S. & Maxwell, S. (eds) *Back from the Brink: Refining the Threatened Species Recovery Process*. Chipping Norton: Surrey Beatty & Sons, pp. 55–62.

BUTT (Butterflies Under Threat Team) (1986). *The Management of Chalk Grassland for Butterflies*. Focus on Nature Conservation no. 17. Peterborough: Nature Conservancy Council.

Canzano, A., New, T. R. & Yen, A. L. (2007). The Eltham copper butterfly, *Paralucia pyrodiscus lucida* Crosby (Lepidoptera: Lycaenidae): local versus state conservation strategies in Victoria. *Victorian Naturalist* **124**, 236–42.

Caughley, G. (1994). Directions in conservation biology. *Journal of Animal Ecology* **63**, 215–44.

Caughley, G., Grice, D., Barker, R. & Brown, B. (1988). The edge of the range. *Journal of Animal Ecology* **57**, 771–85.

Chelmick, D., Hammond, C., Moore, N. & Stubbs, A. (1980). *The Conservation of Dragonflies*. London: Nature Conservancy Council.

Clark, J. A., Hoekstra, J. M., Boersma, P. D. & Kareiva, P. (2002). Improving U.S. Endangered Species Act recovery plans: key findings and recommendations of the SCB recovery plan project. *Conservation Biology* **16**, 1510–19.

Clark, T. W. (1996). Appraising threatened species recovery efforts: practical recommendations. In Stephens, S. & Maxwell, S. (eds) *Back from the Brink: Refining the Threatened Species Recovery Process*. Chipping Norton: Surrey Beatty & Sons, pp. 1–22.

Clarke, C. A. & Sheppard, P. M. (1956). Hand-pairing of butterflies. *Lepidopterists' News* **10**, 47–53.

Clarke, G. & Spier, F. (2003). *A Review of the Conservation Status of Selected Australian Non-marine Invertebrates*. Canberra: Environment Australia/National Heritage Trust.

Cleary, D. F. R. (2004). Assessing the use of butterflies as indicators of logging in Borneo at three taxonomic levels. *Journal of Economic Entomology* **97**, 429–35.

Cochrane, J. F. & Delphey, P. (2002). *Status Assessment and Conservation Guidelines. Dakota skipper,* Hesperia dacotae *(Lepidoptera: Hesperiidae).* Bloomington, MN: United States Fish and Wildlife Service.

Coleman, P. & Coleman, F. (2000). *Local Recovery Plan for the Yellowish Sedge-skipper and Thatching Grass.* Adelaide, South Australia: Delta Environmental Consulting.

Collins, N. M. (1987). *Legislation to Conserve Insects in Europe.* Amateur Entomologist's Society Pamphlet no. 13. London: Amateur Entomologist's Society.

Collins, N. M. & Morris, M. G. (1985). *Threatened Swallowtail Butterflies of the World.* Gland and Cambridge: IUCN.

Commonwealth of Australia (2006). *Threat Abatement Plan to Reduce the Impacts of Tramp Ants on Biodiversity in Australia and its Territories. Plan and Background Document.* Canberra: Department of the Environment and Heritage.

Conrad, K. F., Woiwod, I. P., Parsons, M., Fox, R. & Warren, M. S. (2004). Long-term population trends in widespread British moths. *Journal of Insect Conservation* **8**, 119–36.

Corbet, P. S. (1999). *Dragonflies. Behaviour and Ecology of Odonata.* Colchester: Harley Books.

COSEWIC (Committee on the Status of Endangered Wildlife in Canada) (2003). *Assessment and Status Report on the Dakota skipper* Hesperia dacotae *in Canada.* Ottawa: COSEWIC.

Crone, E. E., Pickering, D. & Schultz, C. B. (2007). Can captive breeding promote recovery of endangered butterflies? An assessment in the face of uncertainty. *Biological Conservation* **139**, 103–12.

Crosby, D. F. (1990). *A Management Plan for the Altona Skipper Butterfly* Hesperilla flavescens flavescens *Waterhouse (Lepidoptera: Hesperiidae).* Technical Report Series no. 98. Victoria: Arthur Rylah Institute for Environmental Research, Department of Conservation, Forests and Lands.

Crozier, L. G. (2004). Field transplants reveal summer constraints on a butterfly range expansion. *Oecologia* **141**, 148–57.

Davies, Z. G., Wilson, R. J., Coles, S. and Thomas, C. D. (2006). Changing habitat associations of a thermally constrained species, the silver-spotted skipper butterfly, in response to climate warming. *Journal of Animal Ecology* **75**, 247–56.

Davis, J. A., Rolls, S. W. & Balla, S. A. (1987). The role of Odonata and aquatic Coleoptera as indicators of environmental quality in wetlands. In Majer, J. D. (ed.) *The Role of Invertebrates in Conservation and Biological Survey.* Department of Conservation and Land Management, Perth, Western Australia, pp. 31–42.

DEH (Department of the Environment and Heritage) (2006). *EPBC Act Policy Statement 1.1. Significant Impact Guidelines. Matters of National Environmental Significance.* Canberra: Commonwealth of Australia.

Dempster, J. P. & Hall, M. L. (1980). An attempt at re-establishing the swallowtail butterfly at Wicken Fen. *Ecological Entomology* **5**, 327–34.

Dennis, R. L. H. & Shreeve, T. G. (1996). *Butterflies on British and Irish Offshore Islands: Ecology and Biogeography.* Oxford: Gem Publishing.

Dennis, R. L. H., Shreeve, T. G. & Van Dyck, H. (2003). Towards a functional resource-based concept for habitat: a butterfly biology viewpoint. *Oikos* **102**, 417–26.

Dennis, R. L. H., Shreeve, T. G. & Van Dyck, H. (2006). Habitats and resources: the need for a resource-based definition to conserve butterflies. *Biodiversity and Conservation* **15**, 1943–68.

Descimon, H. & Napolitano, M. (1992). Genetic management of butterfly populations. In Pavlicek-van Beek, T., Ovaa, A. H. & van der Made, J. G. (eds) *Future of Butterflies in Europe: Strategies for Survival.* Wageningen: Agricultural University, pp. 231–8.

De Whalley, L., De Whalley, B., Green, P., Gammon, N. & Shreeves, W. (2006). Digging scrapes to enhance silver-studded blue *Plebejus argus* habitat at Broadcroft Quarry, Isle of Portland, Dorset, England. *Conservation Evidence* **3**, 39–43.

DEWR (Department of the Environment and Water Resources) (2007). *Developing the Proposed Priority Assessment List of Nominations.* www.environment.gov.au/biodiversity/threatened/ppal-developinh.html (accessed 1 August 2007).

Dirig, R. (1994). Historical notes on wild lupine and the Karner blue butterfly at the Albany Pine Bush, New York. In Andow, D. A., Baker, R. J. & Lane, C. P. (eds) *Karner Blue Butterfly: a Symbol of a Vanishing Landscape.* Miscellaneous Publication 84–1994. St Paul, MN: Minnesota Agricultural Experiment Station, pp. 23–36.

DNRE (Department of Natural Resources and Environment) (2001). *Wetlands. Resource Material for Teachers.* Victoria, Melbourne: Department of Natural Resources and Environment.

Doeg, T. & Reed, J. (1995). Distribution of the endangered Otway stonefly *Eusthenia nothofagi* Zwick (Plecoptera: Eustheniidae) in the Otway Ranges. *Proceedings of the Royal Society of Victoria* **107**, 45–50.

Douglas, F. (2003). Five threatened Victorian sun moths (*Synemon* species). Flora and Fauna Guarantee Act 1988, Action Statement no. 146. Victoria: Department of Sustainability and Environment.

Dover, J. W. & Rowlingson, B. (2005). The western jewel butterfly (*Hypochrysops halyaetus*): factors affecting adult butterfly distribution within native *Banksia* bushland in an urban setting. *Biological Conservation* **122**, 599–609.

Duffey, E. (1977). The reestablishment of the large copper butterfly *Lycaena dispar* (Haw.) *batavus* (Obth.) at Woodwalton Fen National Nature Reserve, Huntingdonshire, England 1969–73. *Biological Conservation* **12**, 143–58.

Duffey, E. & Mason, G. (1970). Some effects of summer floods on Woodwalton Fen in 1968/69. *Entomologist's Gazette* **21**, 23–6.

Dunn, R. R. (2005). Modern insect extinctions, the neglected majority. *Conservation Biology* **19**, 1030–6.

Elkins, R. Fans exterminate 'Hitler' beetle. *The Independent*, 20 August 2006. www.independent.co.uk/news/europe/fans-exterminate-hitler-beetle-412631.html (accessed 5 March 2008).

Ellis, S. (2003). Habitat quality and management for the northern brown argus butterfly *Aricia artaxerxes* (Lepidoptera: Lycaenidae) in North East England. *Biological Conservation* **113**, 285–94.

Emmel, T. C. & Garraway, E. (1990). Ecology and conservation biology of the homerus swallowtail in Jamaica (Lepidoptera: Papilionidae). *Tropical Lepidoptera* **1**, 63–76.

Englund, R. A. (1999). The impacts of introduced poeciliid fish and Odonata on the endemic *Megalagrion* (Odonata) damselflies on Oahu Island, Hawaii. *Journal of Insect Conservation* **3**, 225–43.

Falk, D. A., Millar, C. I. & Olwell, M. (eds) (1996). *Restoring Diversity. Strategies for Reintroduction of Endangered Plants*. Washington, D.C.: Island Press.

Fiedler, P. L. O. & Lavern, R. D. (1996). Selecting reintroduction sites. In Falk, D. A., Millar, C. I. & Olwell, M. (eds). *Restoring Diversity. Strategies for Reintroduction of Endangered Plants*. Washington, D.C.: Island Press, pp. 157–69.

Field, S. A., O'Connor, P. J., Tyre, A. J. & Possingham, H. P. (2007). Making monitoring meaningful. *Austral Ecology* **32**, 485–91.

Fisher, P., Spurr, E. B., Ogilvie, A. C. & Eason, C. T. (2007). Bait consumption and residual concentrations of diphacinone in the Wellington tree weta (*Hemideina crassidens*) (Orthoptera: Anostostomatidae). *New Zealand Journal of Ecology* **31**, 104–10.

Fitzpatrick, U., Murray, T. E., Paxton, R. J. & Brown, M. J. F. (2007). Building on IUCN regional red lists to produce lists of species of conservation priority: a model with Irish bees. *Conservation Biology* **21**, 1324–32.

Forister, M. L. & Shapiro, A. M. (2003). Climatic trends and advancing spring flight of butterflies in lowland California. *Global Change Biology* **9**, 1130–5.

Forister, M. L., Fordyce, J. A. & Shapiro, A. M. (2004). Geological barriers and restricted gene flow in the Holarctic skipper *Hesperia comma* (Hesperiidae). *Molecular Ecology* **13**, 3489–99.

Foster, S. E. & Soluk, D. A. (2006). Protecting more than the wetland: the importance of biased sex ratios and habitat segregation for conservation of Hine's emerald dragonfly, *Somatochlora hineana* Williamson. *Biological Conservation* **127**, 158–66.

Foucart, A. & Lecoq, M. (1998). Major threats to a protected grasshopper, *Prionotropis hystrix rhodanica* (Orthoptera, Pamphagidae, Akicerinae), endemic to southern France. *Journal of Insect Conservation* **2**, 187–93.

Fowles, A. P., Bailey, M. P. & Hale, A. D. (2004). Trends in the recovery of a rosy marsh moth *Coenophila subrosea* (Lepidoptera, Noctuidae) population in response to fire and conservation management on a lowland raised mire. *Journal of Insect Conservation* **8**, 149–58.

Frankham, R. (1995). Effective population size/adult population size ratios in wildlife: a review. *Genetical Research* **66**, 95–107.

Franzen, M. & Nilsson, S. G. (2007). What is the required minimum landscape size for dispersal? *Journal of Animal Ecology* **76**, 1224–30.

Fry, G., Robson, W. & Banham, A. (1992). *Corridors and Barriers to Butterfly Movements in Contrasting Landscapes*. NINA Research Report. Trondheim: Norwegian Institute for Nature Preservation.

Gärdenfors, U. (1996). Application of IUCN Red List categories on a regional scale. In Baillie, J. & Groombridge, B. (eds) *1996 IUCN Red List of Threatened Animals*. Gland: IUCN, pp. 63–6.

Geertsema, H. & Owen, C. R. (2007). Notes on the habitat and adult behaviour of three red-listed *Colophon* spp. (Coleoptera: Lucanidae) of the Cape Floristic Region, South Africa. *Journal of Insect Conservation* **11**, 43–6.

Gerber, L. R. & Hatch, L. T. (2002). Are we recovering? An evaluation of recovery criteria under the U.S. Endangered Species Act. *Ecological Applications* **12**, 668–73.

Gibson, L. & New, T. R. (2007). Problems in studying populations of the golden sun-moth, *Synemon plana* (Lepidoptera: Castniidae) in south eastern Australia. *Journal of Insect Conservation* **11**, 309–13.

Goldstein, P. Z. & De Salle, R. (2003). Calibrating phylogenetic species formation in a threatened insect using DNA from historical specimens. *Molecular Ecology* **12**, 1993–8.

Greenslade, P. (1999). What entomologists think about listing species for protection. In Ponder, W. & Lunney, D. (eds) *The Other 99%. The Conservation and Biodiversity of Invertebrates*. Mosman: Royal Zoological Society of New South Wales, pp. 345–9.

Grill, A., Cleary, D. F. R., Stettner, C., Bräu, M. & Settele, J. (2008). A mowing experiment to evaluate the influence of management on the activity of host ants of *Maculinea* butterflies. *Journal of Insect Conservation* **12**, 617–27.

Grove, S. J. & Stork, N. E. (1999). The conservation of saproxylic insects in tropical forests: a research agenda. *Journal of Insect Conservation* **3**, 67–74.

Grundel, R., Pavlovic, N. B. & Sulzman, C. L. (1998). Habitat use by the endangered Karner blue butterfly in oak woodlands: the influence of canopy cover. *Biological Conservation* **85**, 47–53.

Hanski, I. (1999). *Metapopulation Ecology*. Oxford: Oxford University Press.

Hanski, I. (2005). *The Shrinking World: Ecological Consequences of Habitat Loss*. Oldendorf/Luhe: International Ecology Institute.

Hanski, I. & Gilpin, M. (1991). Metapopulation dynamics: brief history and conceptual domain. *Biological Journal of the Linnean Society* **42**, 3–16.

Hanski, I. & Pöyry, J. (2007). Insect populations in fragmented habitats. In Stewart, A. J. A., New, T. R. & Lewis, O. T. (eds) *Insect Conservation Biology*. Wallingford: CABI, pp. 175–202.

Harrison, S. (1994). Metapopulations and conservation. In Edwards, P. J., May, R. M. & Webb, N. R. (eds) *Large-scale Ecology and Conservation Biology*. Oxford: Blackwell, pp. 111–28.

Haslett, J. R. (1997). *Suggested Additions to the Invertebrate Species listed in Appendix II of the Bern Convention*. Final report to the Council of Europe, Strasbourg.

Hassall, C., Thompson, D. J., French, G. C. & Harvey, I. F. (2007). Historical changes in the phenology of British Odonata are related to climate. *Global Change Biology* **13**, 933–41.

Hauser, C. E., Pople, A. R. & Possingham, H. P. (2006). Should managed populations be monitored every year? *Ecological Adaptations* **16**, 807–19.

Heikkinen, R. K., Luoto, M., Kuussaari, M. & Pöyry, J. (2005). New insights into butterfly-environment relationships using partitioning methods. *Philosophical Transactions of the Royal Society* B **272**, 2203–10.

Henle, K., Davies, K. F., Kleyer, M., Margules, C. & Settele, J. (2004). Predictors of species sensitivity to fragmentation. *Biodiversity and Conservation* **13**, 207–51.

Henning, G. (2001). Notes on butterfly conservation in South Africa and the IUCN Red Data Book categories. *Metamorphosis* **12**, 55–68.

Henning, S. F. & Henning, G. A. (1989). *South African Red Data Book – Butterflies.* South African National Scientific Programmes Report no 158. Pretoria: Council for Scientific and Industrial Research.

Hess, G. R. & Fischer, R. A. (2001). Communicating clearly about conservation corridors. *Landscape and Urban Planning* **55**, 195–208.

Hickling, R., Roy, D. R., Hill, J. K. & Thomas, C. D. (2005). A northward shift of range margins in British Odonata. *Global Change Biology* **11**, 502–6.

Hill, J. K., Thomas, C. D. & Lewis, O. T. (1996). Effects of habitat patch size and isolation on dispersal by *Hesperia comma* butterflies: implications for metapopulation structure. *Journal of Animal Ecology* **65**, 725–35.

Hill, L. & Michaelis, F. B. (1988). *Conservation of Insects and Related Wildlife.* Occasional paper no. 13. Canberra: Australian National Parks and Wildlife Service.

Hingston, A. B. (2007). The potential impact of the large earth bumblebee *Bombus terrestris* (Apidae) on the Australian mainland: lessons from Tasmania. *Victorian Naturalist* **124**, 110–16.

Hochberg, M. E. (2000). What, conserve parasitoids? In Hochberg, M. E. & Ive, A. R. (eds) *Parasitoid Population Biology.* Princeton, NJ: Princeton University Press, pp. 266–77.

Hochberg, M. E., Elmes, G. W., Thomas, J. A. & Clarke, R. T. (1996). Mechanisms of local persistence in coupled host-parasitoid associations: the case model of *Maculinea rebeli* and *Ichneumon eumerus*. *Philosophical Transactions of the Royal Society* B **351**, 1713–24.

Hochkirch, A., Witzenberger, K. A., Teerling, A. & Niemeyer, F. (2007). Translocation of an endangered insect species, the field cricket (*Gryllus campestris* Linnaeus 1758) in northern Germany. *Biodiversity and Conservation* **16**, 3597–607.

Hoffmann, A. A. & Blows, M. R. (1994). Species borders: ecological and evolutionary perspectives. *Trends in Ecology and Evolution* **9**, 223–7.

Holloway, G. J., Dickerson, J. D., Harris, P. W. & Smith, J. (2003a). Dynamics and foraging behaviour of adult hornet robberflies, *Asilus crabroniformis*: implications for conservation management. *Journal of Insect Conservation* **7**, 127–35.

Holloway, G. J., Griffiths, G. H. & Richardson, P. (2003b). Conservation strategy maps: a tool to facilitate biodiversity action planning illustrated using the heath fritillary butterfly. *Journal of Applied Ecology* **40**, 413–21.

Holway, D. A., Lack, L., Suarez, A. V., Tsutsui, N. D. & Case, T. J. (2002). The causes and consequences of ant invasions. *Annual Review of Ecology and Systematics* **33**, 181–233.

Honan, P. (2007a). *Husbandry Manual for the Lord Howe Island Stick Insect.* Melbourne: Melbourne Zoo.

Honan, P. (2007b). The Lord Howe Island stick insect: an example of the benefits of captive management. *Victorian Naturalist* **124**, 258–61.

Horn, D. J. (1991). Potential impact of *Coccinella septempunctata* on endangered Lycaenidae (Lepidoptera) in Northeastern Ohio. In Polagar, L., Chambers, R. J., Dixon, A. F. G. & Hodek, I. (eds) *Behaviour and Impact of Aphidophaga.* The Hague: Academic Publishing, pp. 159–62.

Hutchings, P. A. & Ponder, W. F. (1999). Workshop: criteria for assessing and conserving threatened invertebrates. In Ponder, W. & Lunney, D. (eds) *The Other*

99%. The Conservation and Biodiversity of Invertebrates. Mosman: Royal Zoological Society of New South Wales, pp. 297–315.

IUCN (1987). *IUCN Position Statement on Translocation of Living Organisms.* Gland: IUCN.

IUCN (1994). *IUCN Red List Categories.* IUCN Species Survival Commission. Gland: IUCN.

IUCN (1995). *IUCN/SSC Guidelines for Re-introductions.* Gland: IUCN.

IUCN (2001). *IUCN Red List Categories and Criteria:* Version 3.1. Gland and Cambridge: IUCN/SSC.

IUCN (2003). *Guidelines for Application of IUCN Red List Categories at Regional Level.* Version 3. Gland and Cambridge: IUCN/SSC.

JCCBI (Joint Committee for the Conservation of British Insects) (1971). A code for insect collecting. *Entomologist's Monthly Magazine* **107**, 193–5.

Joseph, L. N., Field, S. A., Wilcox, C. & Possingham, H. P. (2006). Presence–absence versus abundance data for monitoring threatened species. *Conservation Biology* **20**, 1679–87.

KBBRT (Karner Blue Butterfly Recovery Team) (2001). Karner Blue Butterfly Recovery Plan (*Lycaeides melissa samuelis*). Fort Snelling, MN: United States Fish and Wildlife Service.

Key, R. S., Drake, C. M. & Sheppard, D. A. (2000). *Conservation of Invertebrates in England: a Review and Framework.* English Nature Science, Report No 35. Peterborough: English Nature.

Kingsley, K. J. (2002). Population dynamics, resource use, and conservation needs of the Delhi Sands flower-loving fly (*Rhaphiomidas terminalis abdominalis* Cazier) (Diptera: Mydidae), an endangered species. *Journal of Insect Conservation* **6**, 93–101.

Kirby, P. (2001). *Habitat Management for Invertebrates: a Practical Handbook.* Sandy, Bedfordshire: Royal Society for the Protection of Birds.

Kitching, R. L. (1999). Adapting conservation legislation to the idiosyncrasies of the arthropods. In Ponder, W. & Lunney, D. (eds) *The Other 99%. The Conservation and Biodiversity of Invertebrates.* Mosman: Royal Zoological Society of New South Wales, pp. 274–82.

Knisley, C. B., Hill, J. M. & Scherer, A. M. (2005). Translocation of threatened tiger beetle *Cicindela dorsalis dorsalis* (Coleoptera: Cicindelidae) to Sandy Hook, New Jersey. *Annals of the Entomological Society of America* **98**, 552–7.

Koh, L. P., Sodhi, N. S. & Brook, B. W. (2004). Co-extinctions of tropical butterflies and their hostplants. *Biotropica* **36**, 272–4.

Konvicka, M., Maradova, M., Benes, J., Fric, Z. & Kepka, P. (2003). Uphill shifts in distribution of butterflies in the Czech Republic: effects of changing climate on a regional scale. *Global Ecology and Biogeography* **12**, 403–10.

Kudrna, O. (1986). *Butterflies in Europe.* Volume 8. *Aspects of the Conservation of Butterflies in Europe.* Wiesbaden: Aula-Verlag.

Lasswell, H. D. (1971). *A Pre-view of Policy Sciences.* New York: Elsevier.

Lehnert, M. S. (2008). The population biology and ecology of the homerus swallowtail, *Papilio (Pterourus) homerus,* in the Cockpit Country, Jamaica. *Journal of Insect Conservation* **12**, 179–88.

Leigh, J., Briggs, J. & Hartley, W. (1981). *Rare and Threatened Australian Plants.* Special publication no. 7. Canberra: Australian National Parks and Wildlife Service.

Leigh, J., Boden, R. & Briggs, J. (1984). *Extinct and Endangered Plants of Australia*. Sydney: Macmillan Australia.

Leon-Cortes, J. L., Lennon, J. J. & Thomas, C. D. (2003a). Ecological dynamics of extinct species in empty habitat networks. 1. The role of habitat pattern and quantity, stochasticity and dispersal. *Oikos* **102**, 449–64.

Leon-Cortes, J. L., Lennon, J. J. & Thomas, C. D. (2003b). Ecological dynamics of extinct species in empty habitat networks. 2. The role of host plant dynamics. *Oikos* **102**, 465–77.

Leon-Cortes, J. L., Perez-Espinosa, F., Marin, L. & Molina-Martinez, A. (2004). Complex habitat requirements and conservation needs of the only extant Baroniinae swallowtail butterfly. *Animal Conservation* **7**, 241–50.

Lepidopterists' Society (1982). The Lepidopterists' Society statement of committee on collecting policy. *News of the Lepidopterists' Society*, no. 5.

Lewis, O. T. & Basset, Y. (2007). Insect conservation in tropical forests. In Stewart, A. J. A., New, T. R. & Lewis, O. T. (eds) *Insect Conservation Biology*. Wallingford: CABI, pp. 34–56.

Liley, D. (2005). Tree and scrub clearance to enhance habitat for the southern damselfly *Coenagrion mercuriale* at Creech Heath, Dorset, England. *Conservation Evidence* **2**, 131–2.

Lockwood, J. A., Howarth, F. G. & Purcell, M. F. (eds) (2001). *Balancing Nature: Assessing the Impact of Importing Non-native Biological Control Agents (an International Perspective)*. Lanham, MD: Entomological Society of America.

Lovett, G. M., Burns, D. A., Driscoll, C. T. *et al.* (2007). Who needs environmental monitoring? *Frontiers in Ecology and the Environment* **5**, 253–60.

Lundie-Jenkins, G. & Payne, A. (2000). *Recovery plan for the bull oak jewel butterfly* (Hypochrysops piceatus) *1999–2003*. Brisbane: Queensland Parks and Wildlife Service.

Maes, D., Vanreusel, W., Talloen, W. & Van Dyck, H. (2004). Functional conservation units for the endangered Alcon Blue butterfly, *Maculinea alcon* (Lepidoptera: Lycaenidae). *Biological Conservation* **120**, 229–41.

Makibayashi, I. (1996). Recent progress on the conservation effort for *Sasakia charonda* (the great purple emperor) in Ranzanmachi, Saitamo Province. *Decline and Conservation of Butterflies in Japan* **3**, 176–9.

Male, B. (1996). Recovery of Australian threatened species – a national perspective. In Stephens, S. & Maxwell, S. (eds) *Back from the Brink: Refining the Threatened Species Recovery Process*. Chipping Norton: Surrey Beatty & Sons, pp. 23–7.

Martikainen, P. & Kaila, L. (2004). Sampling saproxylic beetles: lessons from a 10-year monitoring study. *Biological Conservation* **120**, 171–81.

Matern, A., Drees, C., Kleinwächter, M. & Assmann, T. (2007). Habitat modeling for the conservation of the endangered ground beetle species *Carabus variolosus* (Coleoptera: Carabidae) in the riparian zone of headwaters. *Biological Conservation* **136**, 618–27.

Mattoni, R. H. T. (1993). The El Segundo Blue, *Euphilotes bernardino allyni* (Shields). In New, T. R. (ed) *Conservation Biology of Lycaenidae (butterflies)*. Occasional Paper of the IUCN Species Survival Commission no 8. Gland: IUCN, pp. 133–4.

Mawdsley, J. R. (2008). Use of simple remote sensing tools to expedite surveys for rare tiger beetles (Insecta: Coleoptera: Cicindelidae). *Journal of Insect Conservation* **12**, 689–93.

Mawdsley, N. A. & Stork, N. E. (1995). Species extinctions in insects: ecological and biogeographical considerations. In Harrington, R. & Stork, N. E. (eds) *Insects in a Changing Environment*. London: Academic Press, pp. 322–69.

Mawson, P. R. & Majer, J. D. (1999). The Western Australian Threatened Species Scientific Committee: lessons from invertebrates. In Ponder, W. & Lunney, D. (eds) *The Other 99%. The Conservation and Biodiversity of Invertebrates*. Mosman: Royal Zoological Society of New South Wales, pp. 369–73.

May, J. E. & Heterick, B. E. (2000). Effects of the coastal brown ant *Pheidole megacephala* (Fabricius) on the ant fauna of the Perth metropolitan region, Western Australia. *Pacific Conservation Biology* **6**, 81–5.

McCormick, B. (2006). *Bogong Moths and Parliament House*. Canberra: Parliament of Australia, Department of Parliamentary Services.

McGuinness, C. (2002). *Threatened Carabid Beetles Recovery Plan* (2002–2007). Wellington: Department of Conservation.

McGuinness, C. A. (2007). Carabid beetle (Coleoptera: Carabidae) conservation in New Zealand. *Journal of Insect Conservation* **11**, 31–41.

McIntire, E. J. B., Schultz, C. B. & Crone, E. E. (2007). Designing a network for butterfly habitat restoration: where individuals, populations and landscapes interact. *Journal of Applied Ecology* **44**, 725–36.

McLaughlin, J. F., Hellmann, J. J., Boggs, C. L. & Ehrlich, P. R. (2002). Climate change hastens population extinctions. *Proceedings of the National Academy of Sciences of the United States of America* **99**, 6070–4.

McQuillan, P. B. & Ek, C. J. (1997). A biogeographical analysis of the Tasmanian endemic Ptunarra brown butterfly, *Oreixenica ptunarra* Couchman (Lepidoptera; Nymphalidae: Satyrinae). *Australian Journal of Zoology* **45**, 21–37.

Meads, M. J. (1994). Translocation of New Zealand's endangered insects as a tool for conservation. In Serena, M. (ed.) *Reintroduction Biology of Australian and New Zealand fauna*. Chipping Norton: Surrey Beatty & Sons, pp. 53–6.

Meggs, J. M., Munks, S. A. & Corkrey, R. (2003). The distribution and habitat characteristics of a threatened lucanid beetle, *Hoplogonus simsoni*, in north-east Tasmania. *Pacific Conservation Biology* **9**, 172–86.

Meggs, J. M., Munks, S. A., Corkrey, R. & Richard, K. (2004). Development and evaluation of predictive habitat models to assist the conservation planning of a threatened lucanid beetle, *Hoplogonus simsoni*, in north-east Tasmania. *Biological Conservation* **118**, 501–11.

Miller, R. M., Rodriguez, J. P., Aniskowicz-Fowler, J. *et al.* (2007). National threatened species listing based on IUCN criteria and regional guidelines: current status and future prospects. *Conservation Biology* **21**, 684–96.

Morgan, M., Knisley, C. B. & Vogler, A. P. (2000). New taxonomic status of the endangered tiger beetle *Cicindela limbata albissima* (Coleoptera: Cicindelidae): evidence from mtDNA. *Annals of the Entomological Society of America* **93**, 1108–15.

Morris, M. G., Thomas, J. A., Ward, L. K. *et al.* (1994). Re-creation of early-successional stages for threatened butterflies – an ecological engineering approach. *Journal of Environmental Management* **42**, 119–35.

Mouquet, N., Belrose, V., Thomas, J. A. *et al.* (2005). Conserving community modules: a case study of the endangered lycaenid butterfly *Maculinea alcon*. *Ecology* **86**, 3160–73.

Munguira, M. L. & Martin, J. (1997). *Action Plan for the Maculinea Butterflies in Europe*. Strasbourg: Council of Europe.

Müller, Z., Jakab, T. & Tóth, A. (2003). Effects of sports fisherman activities on dragonfly assemblages on a Hungarian river floodplain. *Biodiversity and Conservation* **12**, 167–79.

Murphy, D. D. & Weiss, S. B. (1988). A long-term monitoring plan for a threatened butterfly. *Conservation Biology* **2**, 367–74.

Nally, S. (2003). Community involvement in the conservation of endangered purple copper butterfly *Paralucia spinifera* Edwards & Common (Lepidoptera: Lycaenidae). *Records of the South Australian Museum, Monograph series* **7**, 217–24.

New, T. R. (ed.) (1993). *Conservation Biology of Lycaenidae (Butterflies)*. Occasional Paper of the IUCN Species Survival Commission no. 8. Gland: IUCN.

New, T. R. (1995). *Introduction to Invertebrate Conservation Biology*. Oxford: Oxford University Press.

New, T. R. (1998). *Invertebrate Surveys for Conservation*. Oxford: Oxford University Press.

New, T. R. (2005a). *Invertebrate Conservation and Agricultural Ecosystems*. Cambridge: Cambridge University Press.

New, T. R. (2005b). 'Inordinate fondness': a threat to beetles in south east Asia? *Journal of Insect Conservation* **9**, 147–50.

New, T. R. & Britton, D. R. (1997). Refining a recovery plan for an endangered lycaenid butterfly, *Acrodipsas myrmecophila*, in Victoria, Australia. *Journal of Insect Conservation* **1**, 65–72.

New, T. R. & Collins, N. M. (1991). *Swallowtail Butterflies: an Action Plan for their Conservation*. Gland and Cambridge: IUCN.

New, T. R. & Sands, D. P. A. (2002). Narrow-range endemicity and conservation status: interpretations for Australian butterflies. *Invertebrate Systematics* **16**, 665–70.

New, T. R. & Sands, D. P. A. (2003). The listing and de-listing of invertebrate species for conservation in Australia. *Journal of Insect Conservation* **7**, 199–205.

New, T. R. & Sands, D. P. A. (2004). Management of threatened insect species in Australia, with particular reference to butterflies. *Australian Journal of Entomology* **43**, 258–70.

New, T. R., Pyle, R. M., Thomas, J. A., Thomas, C. D. & Hammond, P. C. (1995). Butterfly conservation management. *Annual Review of Entomology* **40**, 57–83.

New, T. R., Van Praagh, B. D. & Yen, A. L. (2000). Fire and the management of habitat quality in an Australian lycaenid butterfly *Paralucia pyrodiscus lucida* Crosby, the Eltham copper. *Metamorphosis* **11**, 154–63.

New, T. R., Gibson, L. A. & Van Praagh, B. D. (2007). The golden sun-moth, *Synemon plana* (Castniidae), on Victoria's remnant southern native grasslands. *Victorian Naturalist* **124**, 254–7.

Nowicki, P., Pepkowska, A., Kudlek, J. *et al.* (2007). From metapopulation theory to conservation recommendations: lessons from spatial occurrence and abundance patterns of *Maculinea* butterflies. *Biological Conservation* **140**, 119–29.

NSWNPWS (New South Wales National Parks and Wildlife Service) (2001). *Bathurst Copper Butterfly (Paralucia spinifera) Recovery Plan.* Hurstville: NSWNPWS.

Oates, M. R. & Warren, M. S. (1990). *A review of Butterfly Introductions in Britain and Ireland.* Godalming: Joint Committee for the Conservation of British Insects and World Wildlife Fund.

O'Dwyer, C. & Attiwill, P. M. (2000). Restoration of a native grassland as habitat for the golden sun moth *Synemon plana* Walker (Lepidoptera: Castniidae) at Mount Piper, Australia. *Restoration Ecology* **8**, 170–4.

Okabe, K. & Goka, K. (2008). Potential impacts on Japanese fauna of canestrinid mites (Acarina: Astigmata) accidentally introduced with pet lucanid beetles from southeast Asia. *Biodiversity and Conservation* 17, 71–81.

Orsak, L. (1992). Saving the world's largest butterfly, Queen Alexandra's birdwing (*Ornithoptera alexandrae*). Five-year management action plan. Durham, California: Scientific Methods.

Ott, J., Schorr, M., Trockur, B. & Lingenfelder, U. (2007). *Species Protection Programme for the Orange-spotted Emerald (Oxygastra curtisii, Insecta: Odonata) in Germany – the example of the River Our population* (in German). Pensoft, Sofia-Moscow: Invertebrate Ecology and Conservation Monographs 3.

Ouin, A., Aviron, S., Dover, J. & Burel, F. (2004). Complementation/supplementation of resources for butterflies in agricultural landscapes. *Agriculture, Ecosystems and Environment* **103**, 473–9.

Parmesan, C. (1996). Climate and species' range. *Nature* **382**, 765–6.

Parmesan, C., Ryrholm, N., Stefanescu, C. *et al.* (1999). Poleward shifts in geographical ranges of butterfly species associated with regional warming. *Nature* **399**, 579–83.

Parmesan, C., Root, T. L. & Willig, M. R. (2000). Impacts of extreme weather and climate on terrestrial biota. *Bulletin of the American Meteorological Society* **81**, 443–50.

Parsons, M. J. (1992). The world's largest butterfly endangered: the ecology, status and conservation of *Ornithoptera alexandrae* (Lepidoptera: Papilionidae). *Tropical Lepidoptera* **3** (Supplement), 35–62.

Parsons, M. J. (1998). *The Butterflies of Papua New Guinea.* London: Academic Press.

Parsons, M. S. (2004). The United Kingdom Biodiversity Action Plan moths – selection, status and progress on conservation. *Journal of Insect Conservation* **8**, 95–107.

Patrick, B. H. & Dugdale, J. S. (2000). *Conservation Status of the New Zealand Lepidoptera.* Science for Conservation no 136. Wellington: Department of Conservation.

Pavlik, B. M. (1996). Defining and measuring success. In Falk, D. A., Millar, C. I. & Olwell, M. (eds) *Restoring Diversity. Strategies for Reintroduction of Endangered Plants.* Washington, D.C.: Island Press, pp. 127–55.

Pearce-Kelly, P., Jones, R., Clarke, D. *et al.* (1998). The captive rearing of threatened Orthoptera: a comparison of the conservation potential and practical considerations of two species' breeding programmes at the Zoological Society of London. *Journal of Insect Conservation* **2**, 201–10.

Pearce-Kelly, P., Morgan, R., Honan, P. *et al.* (2007). The conservation value of insect breeding programmes: rationale, evaluation tools and example programme case studies. In Stewart, A. J. A., New, T. R. & Lewis, O. T. (eds) *Insect Conservation Biology*. Wallingford: CABI, pp. 57–75.

Pearson, D. L., Knisley, C. B. & Kazilek, C. J. (2006). *A Field Guide to the Tiger Beetles of the United States and Canada*. New York: Oxford University Press.

Petanidou, T., Vokou, D. & Margaris, N. S. (1991). *Panaxia quadripunctaria* in the highly touristic Valley of Butterflies (Rhodes, Greece): conservation problems and remedies. *Ambio* **20**, 124–8.

Peters, R. L. & Darling, J. D. S. (1985). The greenhouse effect and nature reserves. *BioScience* **35**, 707–26.

Pollard, E. & Yates, T. J. (1993). *Monitoring Butterflies for Ecology and Conservation*. London: Chapman & Hall.

Powell, A. F. L. A., Busby, W. H. & Kindscher, K. (2007). Status of the regal fritillary (*Speyeria idalia*) and effects of fire management on its abundance in northeastern Kansas, USA. *Journal of Insect Conservation* **11**, 299–308.

Priddel, D., Carlile, N., Humphrey, M., Fellenberg, S. & Hiscox, D. (2002). Rediscovery of the 'extinct' Lord Howe Island stick insect (*Dryococelus australis* Montrouzier) (Phasmatodea) and recommendations for its conservation. *Biodiversity and Conservation* **12**, 1391–403.

Primack, R., Kobori, M. & Mori, S. (2000). Dragonfly pond restoration promotes conservation awareness in Japan. *Conservation Biology* **14**, 1553–4.

Pullin, A. S. (1996). Restoration of butterfly populations in Britain. *Restoration Ecology* **4**, 71–80.

Pullin, A. S., McLean, I. F. G. & Webb, M. R. (1995). Ecology and conservation of *Lycaena dispar*: British and European perspectives. In Pullin, A. S. (ed.) *Ecology and Conservation of Butterflies*. London: Chapman & Hall, pp. 150–64.

Rabinowitz, D., Cairns, S. & Dillon, T. (1986). Seven forms of rarity and their frequency in the flora of the British Isles. In Soulé, M (ed.) *Conservation Biology: the Science of Scarcity and Diversity*. Sunderland, MA: Sinauer, pp. 182–204.

Ravenscroft, N. O. M. (1992). The ecology of the chequered skipper butterfly *Carterocephalus palaemon* Pallas in Scotland. I: microhabitat selection. *Journal of Applied Ecology* **31**, 613–22.

Ravenscroft, N. O. M. (1995). The conservation of *Carterocephalus palaemon* in Scotland. In Pullin, A. S. (ed.) *Ecology and Conservation of Butterflies*. London: Chapman & Hall, pp. 165–79.

RBKC (Royal Borough of Kensington and Chelsea) (1995). The peacock butterfly and other butterflies. *Local Biodiversity Action Plan*. Section 3: *Local Species Action Plans* **6**, 43–5. London: RBKC.

Rout, T. M., Hauser, C. E. & Possingham, H. P. (2007). Minimise long-term loss or maximise short-term gain? Optimal translocation strategies for threatened species. *Ecological Modelling* **201**, 67–74.

Roy, D. B. & Sparks, T. H. (2000). Phenology of British butterflies and climate change. *Global Change Biology* **6**, 407–16.

Samways, M. J. (2003). Threats to the tropical island dragonfly fauna (Odonata) of Mayotte, Comoro archipelago. *Biodiversity and Conservation* **12**, 1785–92.

Samways, M. J. (2005). *Insect Diversity Conservation*. Cambridge: Cambridge University Press.

Samways, M. J., McGeoch, M. A. & New, T. R. (2009). Insect Conservation: approaches and methods. (In preparation.)

Sands, D. P. A. (1999). Conservation status of Lepidoptera: assessment, threatening processes and recovery actions. In Ponder, W. & Lunney, D. (eds) *The Other 99%. The Conservation and Biodiversity of Invertebrates*. Mosman: Royal Zoological Society of New South Wales, pp. 382–7.

Sands, D. P. A. & New, T. R. (2002). *The Action Plan for Australian Butterflies*. Canberra: Environment Australia.

Sands, D. P. A. & New, T. R. (2008). Irregular diapause, apparency and evaluating conservation status; anomalies from the Australian butterflies. *Journal of Insect Conservation* **12**, 81–5.

Sands, D. & Scott, S. (eds) (2002). *Conservation of Birdwing Butterflies*. Chapel Hill, Queensland: SciComEd Pty, Marsden and THECA.

Sands, D. P. A., Scott, S. E. & Moffat, R. (1997). The threatened Richmond birdwing butterfly (*Ornithoptera richmondia* [Gray]): a community conservation project. *Memoirs of Museum Victoria* **56**, 449–53.

Savage, E. J. (2002). Options and techniques for managing Chaffy Saw sedge (*Gahnia filum*) as habitat for the Altona skipper butterfly (*Hesperilla flavescens flavescens*). Melbourne: Parks Victoria Conservation Management paper, Occasional series.

Savignano, D. A. (1994). Benefits to Karner blue butterfly larvae from association with ants. In Andow, D. A., Baker, R. J. & Lane, C. P. (eds) *Karner Blue Butterfly: a Symbol of a Vanishing Landscape*. St Paul, MN: Minnesota Agricultural Experiment Station, Miscellaneous Publication 84–1994, pp. 37–46.

Schtickzelle, N., Turlure, C. & Baguette, M. (2007). Grazing management impacts on the viability of the threatened bog fritillary butterfly *Proclossinia eunomia*. *Biological Conservation* **136**, 651–60.

Schultz, C. B. (2001). Restoring resources for an endangered butterfly. *Journal of Applied Ecology* **38**, 1007–19.

Schultz, C. B. & Crone, E. E. (1998). Burning prairie to restore butterfly habitat: a modeling approach to management tradeoffs for the Fender's blue. *Restoration Ecology* **6**, 244–52.

Schultz, C. B. & Dlugosch, K. (1999). Nectar and hostplant scarcity limit populations of an endangered Oregon butterfly. *Oecologia* **119**, 231–8.

Schultz, C. B. & Hammond, P. C. (2003). Using population viability analysis to develop recovery criteria for endangered insects: case study of the Fender's blue butterfly. *Conservation Biology* **17**, 1372–85.

Schweitzer, D. F. (1994). Recovery goals and methods for Karner blue butterfly populations. In Andow, D. A., Baker, R. J. & Lane, C. P. (eds) *Karner Blue Butterfly: a Symbol of a Vanishing Landscape*. St Paul, MN: Minnesota Agricultural Experiment Station, Miscellaneous Publication 84–1994, pp. 185–93.

Scott, S. (2002). School and community participation in the Richmond Birdwing Conservation Project. In Sands, D. & Scott, S. (eds) *Conservation of Birdwing Butterflies*. Chapel Hill, Queensland: SciComEd Pty, Marsden and THECA, pp. 18–23.

Settele, J., Kuehn, E. & Thomas, J. (eds) (2005). *Studies on the Ecology and Conservation of Butterflies in Europe*. Volume 2. *Species Ecology along a European Gradient*: Maculinea *Butterflies as a Model*. Sofia-Moscow: Pensoft.

Severns, P. (2008). Exotic grass invasion impacts fitness of an endangered prairie butterfly. *Journal of Insect Conservation* **12**, 651–61.

Shaughnessy, J. P. & Cheesman, O. D. (2005). *Wart-Biter Bush-Cricket* (Decticus verrucivorus) *2004*. National Trust Contract Ref U30560. Wallingford: CABI Bioscience.

Shaw, M. R. (1990). Parasitoids of European butterflies and their study. In Kudrna, O. (ed.) *Butterflies of Europe*. Volume 2. *Introduction to Lepidopterology*. Wiesbaden: Aula-Verlag, pp. 449–79.

Shaw, M. R. & Hochberg, M. E. (2001). The neglect of parasitic Hymenoptera in insect conservation strategies: the British fauna as a prime example. *Journal of Insect Conservation* **5**, 253–63.

Sherley, G. H. (1998). *Threatened Weta Recovery Plan*. Threatened Species Recovery Plan No 25. Wellington: Department of Conservation.

Shirt, D. B. (ed.) (1987). *British Red Data Book 2. Insects*. Peterborough: Nature Conservancy Council.

Siepen, G. (2007). School projects. Richmond birdwing corridors for the Noosa Shire. Richmond Birdwing Recovery Network inc., Newsletter Supplement, p. 21. Kenmore.

Sikes, D. S. & Raithel, C. J. (2002). A review of hypotheses of decline of the endangered American burying beetle (Silphidae: *Nicrophorus americanus* Olivier). *Journal of Insect Conservation* **6**, 103–13.

Slater, M. (2007). Creation of a drystone wall to create egg-laying habitats for grizzled skipper *Pyrgus malvae* at Ryton Wood Meadow Butterfly Conservation Reserve, Warwickshire, England. *Conservation Evidence* **4**, 35–40.

Smith, M. N. (2003). *National Stag Beetle Survey 2002*. London: People's Trust for Endangered Species.

Snyder, N. F. R., Derrickson, S. R., Beissinger, S. R. *et al.* (1996). Limitations of captive breeding in endangered species recovery. *Conservation Biology* **10**, 338–48.

Soluk, D. A., Zercher, D. S. & Swisher, B. J. (1998). Preliminary assessment of *Somatochlora hineana* larval habitat and patterns of adult flight over railway lines near Lochport and Lemont, Illinois. Champaign: Illinois Natural History Survey.

Sommers, L. A. & Nye, P. E. (1994). Status, research and management of the Karner blue butterfly in New York. In Andow, D. A., Baker, R. J. & Lane, C. P. (eds) *Karner Blue Butterfly: a Symbol of a Vanishing Landscape*. St Paul, MN: Minnesota Agricultural Experiment Station, Miscellaneous Publication 84–1994, pp. 129–34.

Southwood, T. R. E. & Henderson, P. A. (2000). *Ecological Methods*. Oxford: Blackwell Science.

Sparks, T. H., Dennis, R. L. H., Croxton, P. J. & Cade, M. (2007). Increased migration of Lepidoptera linked to climate change. *European Journal of Entomology* **104**, 139–43.

Speight, M. C. D. (1989). *Saproxylic Invertebrates and their Conservation*. Nature and Environment Series, no. 42. Strasbourg: Council of Europe.

Steencamp, C. & Stein, R. (1999). *The Brenton Blue Saga. A Case Study of South African Biodiversity Conservation.* Parkview: Endangered Wildlife Trust.

Steinbauer, M. J., Yonow, T., Reid, I. A. & Cant, R. (2002). Ecological biogeography of species of *Gelonus*, *Acantholybas* and *Amorbus* in Australia. *Austral Ecology* **27**, 1–25.

Stewart, A. J. A. & New, T. R. (2007). Insect conservation in temperate biomes: issues, progress and prospects. In Stewart, A. J. A., New, T. R. & Lewis, O. T. (eds) *Insect Conservation Biology*. Wallingford: CABI, pp. 1–33.

Stewart, A. J. A., New, T. R. & Lewis, O. T. (eds) (2007). *Insect Conservation Biology*. Wallingford: CABI.

St Louis Zoo (2004). Center for Conservation of the North American Burying Beetle. www:stlzoo.org/wildcareinstitute/americanburyingbeetleinmi/ (accessed 5 November, 2007).

Stork, N. E. & Lyal, C. H. C. (1993). Extinction or 'co-extinction' rates? *Nature* **366**, 307.

Strauss, B. & Biedermann, R. (2005). The use of habitat models in conservation of rare and endangered leafhopper species (Hemiptera: Auchenorrhyncha). *Journal of Insect Conservation* **9**, 245–59.

Stubbs, A. E. (1985). Is there a future for butterfly collecting in Britain? *Proceedings and Transactions of the British Entomological and Natural History Society* **18**, 65–73.

Suhling, F. (1999). Effects of fish on the microdistribution of different larval size groups of *Onychogomphus uncatus* (Odonata: Gomphidae). *Archiv für Hydrobiologie* **144**, 229–44.

Sutter, R. D. (1996). Monitoring. In Falk, D. A., Millar, C. I. & Olwell, M. (eds) *Restoring Diversity. Strategies for Reintroduction of Endangered Plants*. Washington, D.C.: Island Press, pp. 235–64.

Sutton, R. (2006). The effect of cutting grass for butterfly conservation at Witch Lodge Field, Somerset, England. *Conservation Evidence* **3**, 49–51.

Swengel, A. B. & Swengel, S. R. (2007). Benefit of permanent non-fire refugia for Lepidoptera conservation in fire-managed areas. *Journal of Insect Conservation* **11**, 263–79.

Tansy, C. I. (2006). *Hungerford's Crawling Water Beetle (Brychius hungerfordi) Recovery Plan.* Fort Snelling, MN: US Fish and Wildlife Service.

Thomas, C. D. (1995). Ecology and conservation of butterfly metapopulations in the fragmented British landscape. In Pullin, A. S. (ed.) *Ecology and Conservation of Butterflies*. London: Chapman & Hall, pp. 46–63.

Thomas, J. A. (1983). The ecology and conservation of *Lysandra bellargus* (Lepidoptera: Lycaenidae) in Britain. *Journal of Applied Ecology* **20**, 59–83.

Thomas, J. A. (1984). The conservation of butterflies in temperate countries: past efforts and lessons for the future. In Vane-Wright, R. & Ackery, P. (eds) *Biology of Butterflies*. London: Academic Press, pp. 333–53.

Thomas, J. A. (1995a). The ecology and conservation of *Maculinea arion* and other European species of large blue butterfly. In Pullin, A. S. (ed.) *Ecology and Conservation of Butterflies*. London: Chapman & Hall, pp. 180–97.

Thomas, J. A. (1995b). Why small cold-blooded insects pose different conservation problems to birds in modern landscapes. *Ibis* **137** (suppl.), 112–19.

Thomas, J. A. & Elmes, G. W. (1992). The ecology and conservation of *Maculinea* butterflies and their ichneumon parasites. In Pavlicek-van Beek, T., Ovaa, A. H. & van der Made, J. G. (eds) *Future of Butterflies in Europe: Strategies for Survival*. Wageningen: Agricultural University, pp. 116–23.

Toledo Zoo (2002). *Propagation Manual for the Karner Blue Butterfly, Lycaeides melissa samuelis*. Toledo, OH: Toledo Zoo, Department of Conservation and Research.

USFWS (United States Fish and Wildlife Service) (1993). *Northern Beach Tiger Beetle (Cicindela dorsalis dorsalis) Recovery Plan*. Hadley, MA: USFWS.

USFWS (United States Fish and Wildlife Service) (1997). *Recovery Plan for Mitchells Satyr Butterfly (Neonympha mitchellii mitchellii French)*. Fort Snelling, MN: USFWS.

USFWS (United States Fish and Wildlife Service) (1999). *Schaus Swallowtail Butterfly, Heraclides aristodemus ponceanus*. In *Multi-Species Recovery Plan for South Florida*. USFWS South Florida Ecological Services Office.

USFWS (United States Fish and Wildlife Service) (2001). *Hine's Emerald Dragonfly (Somatochlora hineana Williamson) Recovery Plan*. Fort Snelling, MN: USFWS.

USFWS (United States Fish and Wildlife Service) (2005). *Dakota Skipper Conservation Guidelines, Hesperia dacotae (Skinner) (Lepidoptera: Hesperiidae)*. USFWS, Twin Cities Field Office. Available at http://midwest.fws.gov/endangered/insects/dask-cons-guid.pdf.

van Swaay, C. & Warren, M. S. (1999). *Red Data Book of European Butterflies (Rhopalocera)*. Nature and Environment no. 99. Strasbourg: Council of Europe.

van Swaay, C. & Warren, M. S. (2003). *Prime Butterfly Areas in Europe: Priority Sites for Conservation*. Wageningen: National Reference Centre for Agriculture.

van Tol, J. & Verdonk, M. J. (1988). *The Protection of Dragonflies (Odonata) and their Biotopes*. Nature and Environment Series no. 38. Strasbourg: Council of Europe.

Vane-Wright, R. I., Humphries, C. J. & Williams, P. H. (1991). What to protect? Systematics and the agony of choice. *Biological Conservation* **55**, 235–54.

Vaughan, P. J. (1988). Management plan for the Eltham copper butterfly (*Paralucia pyrodiscus lucida* Crosby) (Lepidoptera: Lycaenidae). Technical Report Series no. 79. Melbourne: Arthur Rylah Institute for Environmental Research, Department of Conservation, Forests and Lands.

Vogler, A. P. & De Salle, R. (1994). Diagnosing units of conservation management. *Conservation Biology* **8**, 354–63.

Vogler, A. P., Knisley, C. B., Glueck, S. B., Hill, J. M. & De Salle, R. (1993). Using molecular and ecological data to diagnose endangered populations of the Puritan tiger beetle, *Cicindela puritana*. *Molecular Ecology* **2**, 375–83.

Walters, A. C. (2006). Invasion of Argentine ants (Hymenoptera: Formicidae) in South Australia: impacts on community composition and abundance of invertebrates in urban parklands. *Austral Ecology* **31**, 567–76.

Warren, M. S., Hill, J. K., Thomas J. A. *et al.* (2001). Rapid responses of British butterflies to opposing forces of climate and habitat change. *Nature* **414**, 65–9.

Wells, S. M., Pyle, R. M. & Collins, N. M. (1983). *The IUCN Invertebrate Red Data Book*. Gland and Cambridge: IUCN.

Williams, S. (1996). Community involvement in the species recovery process: insights into successful partnerships. In Stephens, S. & Maxwell, S. (eds) *Back from the Brink: Refining the Threatened Species Recovery Process*. Chipping Norton: Surrey Beatty & Sons, pp. 87–96.

Wilson, R.J., Davies, Z.G. & Thomas, C.D. (2007). Insects and climate change: process, patterns and implications for conservation. In Stewart, A.J.A., New, T.R. & Lewis, O.T. (eds) *Insect Conservation Biology*. Wallingford: CABI, pp. 245–79.

Windsor, D.A. (1995). Equal rights for parasites. *Conservation Biology* **9**, 1–2.

Wynhoff, I. (1998). Lessons from the reintroduction of *Maculinea teleius* and *M. nausithous* in the Netherlands. *Journal of Insect Conservation* **2**, 47–57.

Yen, A.L. & Butcher, R.J. (1997). *An Overview of the Conservation Status of Non-marine Invertebrates in Australia*. Canberra: Environment Australia.

Zborowski, P. & Edwards, T. (2007). *A Guide to Australian Moths*. Melbourne: CSIRO Publishing.

Index